100 Modern Reagents

100
Modern
Reagents

Editor: N. S. Simpkins

The Royal Society of Chemistry

This book was produced for the Royal Society of Chemistry by

Moss Publishing Ltd
9 Heathcoat Building
Highfields Science Park
Nottingham NG7 2QJ

Tel. 0602 257979
FAX 0602 254261

Lynda Richards   Director
Jeanette Eldridge   Assistant Editor
Angela Bluhm   Production Supervisor
Denise Evans   Production Assistant

Data Sheet Design   Iwan Thomas (BKT Information Services)
Structure Generation   Steve Stokes (Nottingham University)
Cover Design   John Tanton (RSC)
Product Manager   Michael Hannant (RSC)

Printed by Audiovisual Services, University of Technology, Loughborough, Leics.

ISBN 0-85186-893-2

Reasonable care has been taken in the preparation of *100 Modern Reagents* but the
RSC does not accept liability for the consequences of any errors or omissions.
Inclusion of an item in *100 Modern Reagents* does not imply endorsement by the
RSC of the content of the original document to which the item refers.

Foreword

One of the primary requirements of modern organic synthesis is the highly selective chemical transformation of complex molecules to desirable target compounds. The difficulties in efficiently conducting such selective reactions have resulted in the development of such a vast armoury of reagents that it is ever more difficult to find the ideal reagent for a particular problem. The need for a lab companion that could act both as a source-book, and as a point of access to the primary literature, has resulted in this book, *100 Modern Reagents*.

The book brings together a selection of 100 important reagents which have found significant application in organic synthesis. *100 Modern Reagents* features both well-known and less-familiar reagents which have been identified in *Methods in Organic Synthesis* during the last five years. The selection reflects reagents which have found widespread use, with emphasis on those which conduct important transformations, or which show interesting and varied reactivity.

Chemical and physical data are provided for each reagent, along with up-to-date information on safety precautions, literature preparations or commercial availability. Each page of data is accompanied by a page of reaction schemes illustrating recent applications of the particular reagent, along with literature references. *100 Modern Reagents* will prove invaluable to practising synthetic chemists in both industry and academia, who are looking for new reagents in synthesis or new uses for more familiar ones.

Contents

Sources

CAS registry numbers, CAS names, physical data, safety and handling information (including flash point (Fp) figures) and brief details on reactions, reviews and preparations were compiled with the aid of the texts listed below.[1-5]

Additional data and details on availability were also obtained from catalogues of chemical suppliers[6-9] and have been presented in the form: Supplier, state, purity and cost. The cost is given as p, £, ££, or £££, indicating a price per gram of less than 50p, 50p-£1.50, £1.50-£5.00, and more than £5.00, respectively, in order to allow general comparisons to be made.

1. *Chemical Abstracts.*

2. *Handbook of Chemistry and Physics,* 60th edn,
   ed. R. C. Weast, CRC Press, Boca Raton, Florida, 1979-1980.

3. *Hazards in the Chemical Laboratory,* 3rd edn,
   ed. L. Bretherick, RSC, London, 1981.

4. *Chemical Safety Data Sheets, Vol. 1: Solvents,* RSC, London, 1989.

5. *Advanced Organic Chemistry,* 2nd edn,
   J. March, McGraw International, Japan, 1983.

6. Aldrich Chemical Co. Ltd, Catalogue 1988-1989.

7. Johnson Matthey Chemicals, Catalogue 1986-1987.

8. Lancaster Synthesis Ltd, Catalogue 1989-1990.

9. Sigma Chemical Co. Ltd, Catalogue, 1989.

1. Acetyl hypofluorite

CAS Registry Number	78948-09-1
CAS Name	Acetic acid, anhydride with hypofluorous acid
Molecular Formula	AcOF
Molecular Weight	78
Boiling Point	Not available.
Melting Point	$-96^{\circ}C$
Density	Not available.
Refractive Index	Not available.
Safety and Handling	Limited stability in liquid state. May explode. See *Chem. Eng. News*, 1985, **63**(*7*), 2. Vapour has 2 h half-life at room temp. Physical properties discussed, see *J. Am. Chem. Soc.*, 1985, **107**(*23*), 6515-6518.
Reactions	Fluorination. Oxygenation.
Availability	Not commercially available.
Preparation	Prepared by bubbling F_2 through suspensions of NaF, NaOAc or NaO_2CCF_3 in 1:9 AcOH:CFCl₃. See S. Rozen *et al.*, *J. Chem. Soc., Chem. Commun.*, 1981, (*10*), 443-444.
Other Preparations	Also prepared by passage of 5% v/v F_2-O_2 through solid KOAc.2HOAc and trapped as slightly yellow liquid at -78°. See also: *Synth. Commun.*, 1984, **14**(*1*), 45; *Carbohydrates*, 1984, **24**(*4*), 477-484; *J. Fluor. Chem.*, 1984, (*24*), 477.

Stereoselective fluorination of carbohydrates

M. J. Adam*, B. D. Pate, J.-R. Nesser, L. D. Hall, *Carbohydr. Res.*, 1983, **124**(2), 215-224

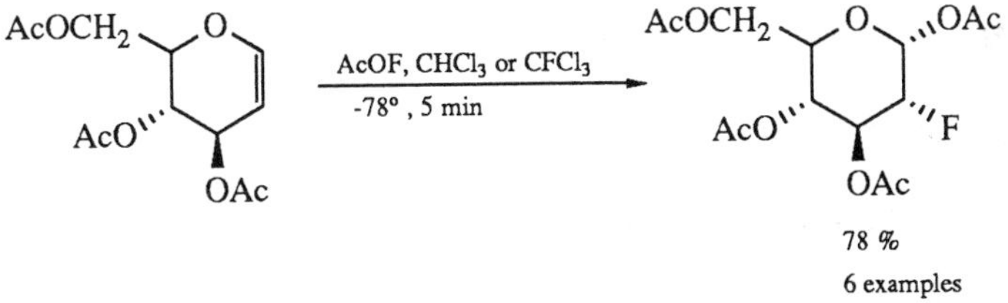
AcOCH2
O
AcO
OAc
AcOF, CHCl3 or CFCl3
-78° , 5 min
AcOCH2
O
OAc
AcO
F
OAc
78 %
6 examples

Direct fluorination of lithium enolates with acetyl hypofluorite

S. Rozen*, M. Brand, *Synthesis*, 1985, (6/7), 665-667

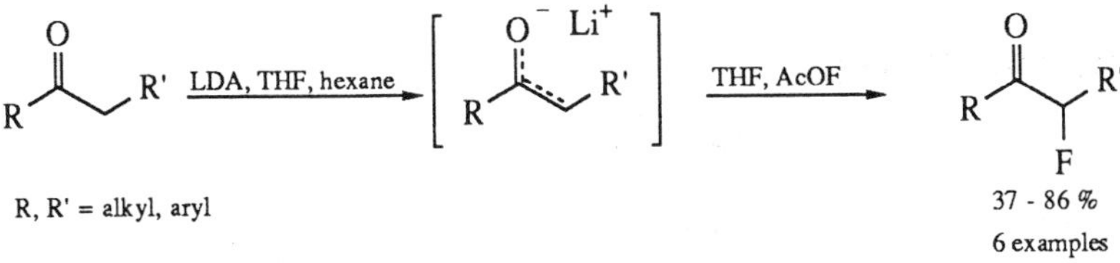
O
R
R'
LDA, THF, hexane
O
Li+
R
R'
THF, AcOF
O
R
R'
F
R, R' = alkyl, aryl
37 - 86 %
6 examples

Chlorination, bromination and oxygenation of pyridines with acetyl hypofluorite

S. Rozen*, D. Heber, *J. Org. Chem.*, 1988, **53**(5), 1123-1125

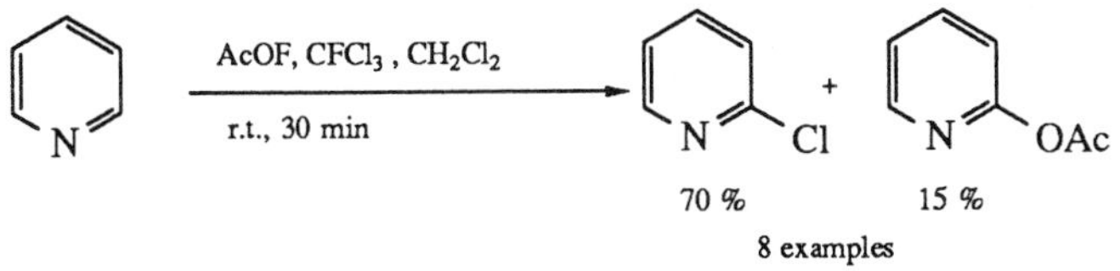
N
AcOF, CFCl3 , CH2Cl2
r.t., 30 min
N
Cl
+
N
OAc
70 %
15 %
8 examples

2. Allyltributyltin

CAS Registry Number   24850-33-7

CAS Name   Stannane, tributyl-2-propenyl-

Molecular Formula $[Me(CH_2)_3]_3SnCH_2CH=CH_2$

Molecular Weight   331.11

Boiling Point   88-92°C/0.2 mmHg

Melting Point   Not available.

Density   1.068 kg/m^3

Refractive Index   1.4846

Safety and Handling   Irritant.

Fp 110°C

Reactions   Allylation.
Reviews: *Chem. Rev.*, 1960, 459; *Chem. Ind. (London)*, 1972, 490; *Aldrichim. Acta*, 1987, **20**(2), 45-49.

Availability   Aldrich: 97%, £££.

Alkenyl-substituted tetrahydrofurans via successive intra- and intermolecular radical reactions

O. Moriya*, M. Kakihana, Y. Urata, T. Sugizaki, T. Kageyama, Y. Ueno, T. Endo, *J. Chem. Soc., Chem. Commun.*, 1985, (20), 1401-1402

R^1 = alkyl, aryl R^2 = H, CH_2OH 40 -70 %   6 examples   25 - 49 %
$R^1 R^2$ = cycloalkyl R^3, R^4 = H, Me

Preparation of allylazetidinones and methylallylazetidinones

H. Fliri*, C.-P. Mak*, *J. Org. Chem.*, 1985, 50(19), 3438-3442

68 %
5 examples

Organotin reactions with isoquinolines. Dibenzoquinolizinone synthesis

R. Yamaguchi*, A. Otsuji, K. Utimoto, *J. Am. Chem. Soc.*, 1988, 110(7), 2186-2187

68 - 94 %   76 - 84 %
6 examples   4 examples

3. Allyltrimethylsilane

CAS Registry Number   762-72-1

CAS Name   Silane, trimethyl-2-propenyl-

Molecular Formula $H_2C=CHCH_2SiMe_3$

Molecular Weight   114.27

Boiling Point   85°C / 737 mm Hg

Melting Point   Not available.

Density   0.719 kg/m^3

Refractive Index   1.4056

Safety and Handling   Highly flammable. Irritant.

Reactions   Allylation. Alkylation.
Reviews: *Pure Appl. Chem.*, 1982, **54**, 1; *Acc. Chem. Res.*, 1988, **21**(5), 200-206.

Availability   Aldrich: 99%, £.

Lancaster Synthesis: 99+%, £, bulk prices available.

Sigma: £.

Alkylation of aldehydes with etherification by dialkoxydichlorotitanium-allyltrimethylsilane. An asymmetric variation of the Sakurai reaction

R. Imwinkel, D. Seebach*, *Angew. Chem. (Int. Ed. Engl.)*, 1985, **24**(9), 765-766

RCHO
i) $Cl_2TiR'_2$, -75°, 30 min.
ii) Me_3Si ⌢, CH_2Cl_2, -75° - r.t.

R = alkyl, aryl
R'= alkoxy, chiral alkoxy

OR'
R ⌢
42 - 95 %
< 91.5 % de

Me_3SiI

OH
R ⌢
75 %
78 - 80 % ee
6 examples

Stereoselective ribofuranosylation. Reaction of ribofuranosyl fluorides with silyl enol ethers and allyltrimethylsilane

Y. Araki, N. Kobayashi, Y. Ishido, J. Nagasawa, *Carbohydr. Res.*, 1987, **171**, 125-139

BnO
O
OH
MeO OMe

$F_3CCHFCF_2NEt_2$
CH_2Cl_2, r.t., 45 min.
23 examples

BnO
O
F
MeO OMe
50 % β–F
35 % α–F

[β–F] ⌢ $SiMe_3$, Et_2O
CH_2Cl_2, r.t., 10 min

BnO
O
MeO OMe
α–
84.2 %

Trimethylsilyl triflate-catalyzed addition of allylsilanes to nitrones

P. G. M. Wutts*, Y.-W. Jung, *J. Org. Chem.*, 1988, **53**(9), 1957-1965

Ph
‖
O^- N^+ Me

Me_3Si ⌢, Me_3SiOTf
CH_2Cl_2, r.t., 36 h

Ph
Me N OH

+

Ph
MeN O $SiMe_3$

90 %
>50:1 mixture
10 examples

4. Aluminium chloride

CAS Registry Number   7446-70-0, 7784-13-6 (hexahydrate)

CAS Name   Aluminum chloride (AlCl$_3$)

Molecular Formula   AlCl$_3$

Molecular Weight   133.34, 241.43 (hexahydrate)

Boiling Point   Not available.

Melting Point   190°C (2.5 atm), 100°C(dec.) (hexahydrate)

Density   2.44 kg/m^3, 2.398 kg/m^3 (hexahydrate)

Refractive Index   1.6 (hexahydrate)

Safety and Handling   Corrosive. Irritant. Violently decomposed by water.

Reactions   Reduction. Lewis acid catalyst. Friedel-Crafts catalyst. Ring enlargement.

Availability   Aldrich: anhydrous, powder and granules, 99.99%, p; anhydrous, 99%, p; anhydrous, p; 1M in nitrobenzene, under N$_2$ in Sure/Seal™, £; hexahydrate, 99.99%, £; hexahydrate, 99%, £.

Johnson Matthey: hexahydrate, crystalline, Spec-pure®, p; hydrate, 97% p; anhydrous, p.

Sigma: hexahydrate, p.

Facile reduction of sulphoxides to sulphides using an aluminium chloride-sodium iodide system

M. V. Bhatt*, J. R. Babu, *Indian J. Chem., Sect. B*, 1988, **27**(*3*), 259-260

$$R^1 - \overset{\overset{\text{O}}{\|}}{S} - R^2 \xrightarrow[\substack{\text{MeCN, r.t. - reflux,} \\ 0.5 - 15 \text{ h}}]{\text{AlCl}_3 \text{, NaI}} R^1 SR^2$$

R^1, R^2 = alkyl, aryl

65 - 92.5 %
8 examples

Regioselective Friedel-Crafts acylation of ethyl indole-2-carboxylate with acyl chlorides or acid anhydrides

Y. Murakami*, M. Tani, K. Tanaka, Y. Yokoyama, *Chem. Pharm. Bull.*, 1988, **36**(*6*), 2023-2035

AlCl₃ , ClCH₂CH₂Cl
p-CH₃OC₆H₄COCl
reflux, 1 h

69.7 %
16 examples

Ring expansion of cyclic ketones to α-phenylthio ketones

S. Kim*, J. H. Park, *Chem. Lett.*, 1988, (*8*), 1323-1324

-80°, 1 h

AlCl₃ , CH₂Cl₂
0°, 30 min

85 %

74 %
7 examples

8

5. Aluminium iodide

CAS Registry Number	7784-23-8
CAS Name	Aluminum iodide (AlI_3)
Molecular Formula	AlI_3
Molecular Weight	407.70, 515.79 (hexahydrate)
Boiling Point	360°C
Melting Point	191°C, 185°C (dec.) (hexahydrate)
Density	3.980 kg/m^3
Refractive Index	Not available.
Safety and Handling	Corrosive. Moisture sensitive.
Reactions	Reduction. Halide exchange. Ether cleavage.
Availability	Aldrich: anhydrous, 95%, £.

An improvement of the aluminium iodide method for ether cleavage: catalysis by quaternary ammonium iodides

S. Andersson, *Synthesis*, 1985, (4), 437-439

$$ArOR \xrightarrow[\text{reflux, 0.3 - 14 h}]{\substack{AlI_3 \text{ , } Bu_4N^+I^- \\ \text{cyclohexane or } C_6H_6 \text{ , } H_2O}} ArOH + RI$$

R = alkyl

14 - 98 %
11 examples

Reduction of sulphonyl chlorides and sulphoxides with aluminium iodide

J. R. Babu, M. V. Bhatt*, *Tetrahedron Lett.*, 1986, **27**(9), 1073-1074

$$RSO_2Cl \xrightarrow[\text{MeCN, r.t.-reflux}]{AlI_3 \text{ , } 0.5 - 55 \text{ h,}} RSSR$$

81 - 95 %
7 examples

$$R-\overset{\overset{\displaystyle O}{\|}}{S}-R \longrightarrow RSR$$

R = alkyl, aryl

68 - 84 %
3 examples

Preparation of alkyl iodides from alkyl chlorides using aluminium triiodide

F. J. Arnaiz, J. M. Bustillo, *An. Quim., Ser. C.*, 1986, **82**(3), 270-271

$$MeCCl_3 \xrightarrow[\text{ii) } H_2O]{\text{i) } AlI_3 \text{ , } CS_2 \text{ ,10 min}} MeCI_3$$

75 - 89 %
12 examples

6. Aluminium isopropoxide

CAS Registry Number   555-31-7

CAS Name   2-Propanol, aluminum salt

Molecular Formula $(Me_2CHO)_3Al$

Molecular Weight   204.25

Boiling Point   140.5°C

Melting Point   138-142°C (99.99+%), 118°C (98+%)

Density   1.035 kg/m^3

Refractive Index   Not available.

Safety and Handling   Corrosive. Moisture sensitive. Flammable. Irritant.

Reactions   Reduction of carbonyl compounds. Ring cleavage. Reviews: *Org. React.*, 1944, **2**, 178; *Org. React.*, 1951, **6**, 207.

Suppliers   Aldrich: 99.99+% (purity based on metals analysis), £; 98+%, p.

Lancaster Synthesis: 98+%, p.

Sigma: p.

Preparation of ethers from aluminium alkoxides and alkyl halides in DMF

L. Lompa-Krzymien, L. C. Leitch, *Polish J. Chem.*, 1983, 57(4-6), 629-630

$$Al(OR)_3 \quad + \quad R'X \xrightarrow[\text{2 days}]{\text{DMF, reflux}} ROR'$$

R,R' = alkyl
X = halide

20 - 80 %
10 examples

Regioselective reduction of pyrimidinones with zirconium or aluminium isopropoxides

T. Hoseggen, F. Rise, K. Undheim*, *J. Chem. Soc., Perkin Trans. 1*, 1986, (5), 849-850

Zr(OPr-i)$_4$ or Al(OPr-i)$_3$

i-PrOH, 90°, 2 days

R = H, halide

20 - 91 %
5 examples

Regio- and chemoselective ring opening of epoxides with trimethylsilyl azide/aluminium isopropoxide

M. Emziane, P. Lhoste, D. Sinou*, *Synthesis*, 1988, (7), 541-544

Me$_3$SiN$_3$ (1.5 equiv.)

Al(OPr-i)$_3$ (0.1 equiv.)

CH$_2$Cl$_2$

R^1 = alkyl, allyl, substit. Me
R^2 = H
or R^{1}R^2 = (CH$_2$)$_n$, n = 3,4,5

59 - 93 %

7. Benzoyl peroxide

CAS Registry Number   94-36-0

CAS Name   Peroxide, dibenzoyl

Molecular Formula $(PhCO)_2O_2$

Molecular Weight   242.23

Boiling Point   Not available.

Melting Point   104-106°C (dec.) (97%), 105°C (70%)

Density   Not available.

Refractive Index   1.545

Safety and Handling   Oxidizer. Irritating to eyes, skin and respiratory system. Found to have skin tumour promoting activity: *Science*, 1981, **213**, 1023. Extreme risk of explosion by shock, friction, fire or other source of ignition.

Reactions   Initiator. Curing agent. Cross-linking agent. Oxidation. Benzoylation.

Availability   Aldrich: 97%, p; 70% (remainder water), p.

Oxidation of naphthols with benzoyl peroxide

T. Matsumoto*, S. Imai*, N. Yamamoto, *Bull. Chem. Soc. Jpn.*, 1988, **61**(*3*), 911-919

$$(PhCO)_2O_2 \ (0.95 \ equiv.)$$
$$CH_2Cl_2 \ , \ r.t., \ 24 \ h$$

52 %
4 examples

Regioselective benzoylation of α-glycols

A. M. Pautard, S. A. Evans, Jr.*, *J. Org. Chem.*, 1988, **53**(*10*), 2300-2303

$$HOCHR\text{-}CH_2OH \xrightarrow[\text{CH}_2\text{Cl}_2 \ , \ \text{Ar, r.t.,1 h}]{\text{Ph}_3\text{P, (PhCO)}_2\text{O}_2} PhCO_2CHR\text{-}CH_2OH$$

R = Me, Ph

80 %

Conversion of alcohols to methylthiomethyl ethers

J. C. Medina, M. Salomon, K. S. Kyler*, *Tetrahedron Lett.*, 1988, **29**(*31*), 3773-3776

i) Me$_2$S, MeCN, O°
ii) (PhCO)$_2$O$_2$, MeCN, 2 h

MTM = methylthiomethyl

88 %
10 examples

8. Benzyltrimethylammonium dichloroiodate

CAS Registry Number   114971-52-7

CAS Name   Benzenemethanaminium, N,N,N-trimethyl-, dichloroiodate(1-)

Molecular Formula $PhCH_2NMe_3ICl_2$

Molecular Weight   348.05

Boiling Point   Not available.

Melting Point   125°C

Density   Not available.

Refractive Index   Not available.

Safety and Handling   Not available.

Reactions   Iodination. Chlorination.

Availability   Not commercially available.

Preparation   Prepared by reaction of ICl in CH_2Cl_2 and $PhCH_2NMe_3Cl$ in H_2O at room temp. as stable brilliant yellow needles: *Chem. Lett.*, 1987, 2109; *Bull. Chem. Soc. Japan*, 1988, **61**, 600-602.

Other Preparations   *Chem. Lett.*, 1988, 795-798.

Bromination and iodination of aryl amines with benzyltrimethylammonium tribromide and dichloroiodate

S. Kajigaeshi*, T. Kakinami, K. Inoue, M. Kondo, H. Nakamura, M. Fujikawa, T. Okamoto, *Bull. Chem. Soc. Jpn.*, 1988, **61**(2), 597-599, 600-602

$$H_2N-\langle\ \rangle \xrightarrow[\text{MeOH, r.t., 30 min}]{PhCH_2Me_3N^+ICl_2^-,\ CaCO_3,\ CH_2Cl_2} H_2N-\langle\ \rangle-I$$

94 %
25 examples

Iodination of aromatic ethers

S. Kajigaeshi*, T. Kakinami, M. Moriwaki, M. Watanabe, S. Fujisaki, T. Okamoto, *Chem. Lett.*, 1988, (5), 795-798

$$R^1O-\langle\ \rangle^{R^2} \xrightarrow[\text{ZnCl}_2,\ \text{AcOH, r.t.}]{PhCH_2Me_3N^+ICl_2^-} R^1O-\langle\ \rangle^{R^2}_{I_n} \quad n = 1,2$$

R^1, R^2 = alkyl, aryl, H

87 - 98 %
20 examples

α-Chlorination of aromatic acetyl derivatives

S. Kajigaeshi*, T. Kakinami, M. Moriwaki, S. Fujisaki, K. Maeno, T. Okamoto, *Synthesis*, 1988, (7), 545-546

$$Ar\overset{O}{\underset{}{\underset{}{\|}}}Me \xrightarrow[\text{reflux, 3 - 10 h}]{\substack{PhCH_2Me_3N^+ICl_2^- \\ ClCH_2CH_2Cl, MeOH}} Ar\overset{O}{\underset{}{\|}}CH_2Cl$$

Ar = Ph, substit. Ph, 2-napthyl
2-thienyl

66 - 99 %
13 examples

9. Bismuth(III) chloride

CAS Registry Number	7787-60-2
CAS Name	Bismuthine, trichloro-
Molecular Formula	$BiCl_3$
Molecular Weight	315.34
Boiling Point	447°C
Melting Point	230-232°C
Density	4.750 kg/m^3
Refractive Index	Not available.
Safety and Handling	Corrosive. Moisture sensitive.
Reactions	Catalyst for allylation. Addition. Catalyst for aldol reaction.
Availability	Aldrich: 99.999%, £; anhydrous, 98+%, p.
	Johnson Matthey: 99.999%, ££; crystalline, Specpure®, £££; 98%, p.

Grignard-type addition of allyl halides to aldehydes

M. Wada, H. Ohki, K. Akiba*, *Tetrahedron Lett.*, 1986, **27**(*39*), 4771-4774

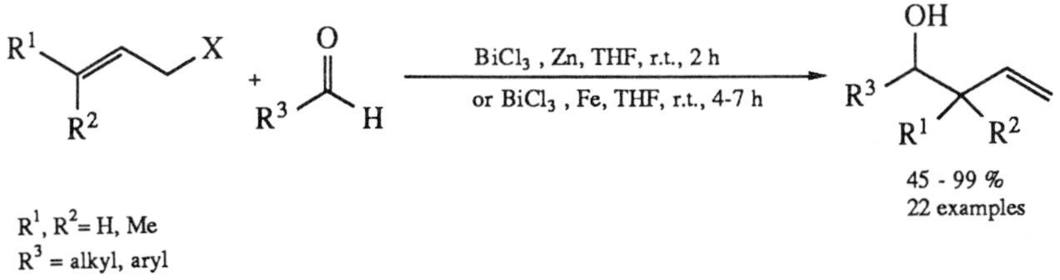
BiCl3 , Zn, THF, r.t., 2 h
or BiCl3 , Fe, THF, r.t., 4-7 h
45 - 99 %
22 examples
R1, R2 = H, Me
R3 = alkyl, aryl

Bismuth(III) chloride-aluminium promoted allylation of aldehydes to homoallyl alcohols

M. Wada*, H. Ohki, K. Akiba*, *J. Chem. Soc., Chem. Commun.*, 1987, (*10*), 708-709

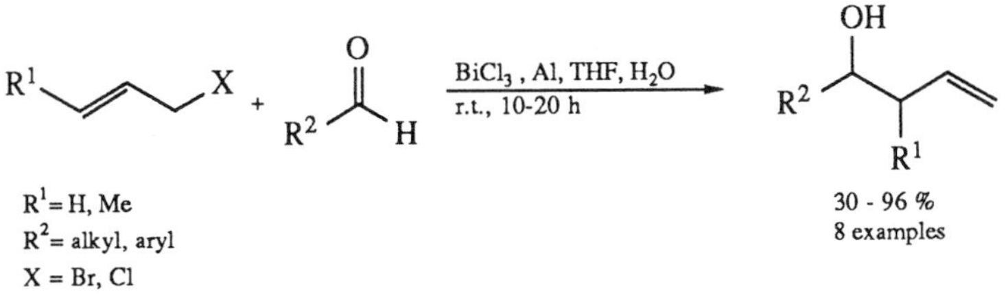
BiCl3 , Al, THF, H2O
r.t., 10-20 h
30 - 96 %
8 examples
R1 = H, Me
R2 = alkyl, aryl
X = Br, Cl

Bismuth trichloride as an efficient catalyst in the aldol reaction

H. Ohki, M. Wada*, K. Akiba*, *Tetrahedron Lett.*, 1988, **29**(*37*), 4719-4722

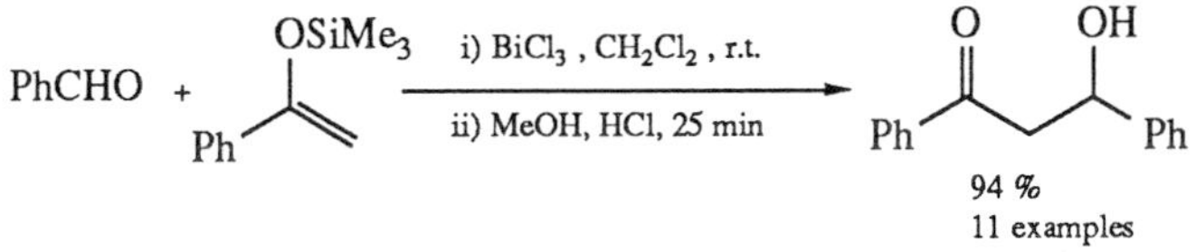
PhCHO
OSiMe3
i) BiCl3 , CH2Cl2 , r.t.
ii) MeOH, HCl, 25 min
94 %
11 examples

10. Bis(tributyltin) oxide

CAS Registry Number	56-35-9
CAS Name	Distannoxane, hexabutyl-
Molecular Formula	$\{[Me(CH_2)_3]_3Sn\}_2O$
Molecular Weight	596.08
Boiling Point	180°C/2 mmHg
Melting Point	Not available.
Density	1.170 kg/m^3
Refractive Index	1.4864
Safety and Handling	Corrosive. Toxic. Keep cold.
Reactions	Oxidizing agent with bromine: *Tetrahedron Lett.*, 1977, 2413; 1978, 1277. Reviews: *Chem. Rev.*, 1960, 459; *Chem. Ind. (London)*, 1972, 490; *Synthesis*, 1969, 56.
Availability	Aldrich: 96%, p. Lancaster Synthesis: 96%, p, bulk prices available.

An effective method for removal of internucleotidic phenylthio groups from fully protected oligonucleotides

M. Sekine*, H. Tanimura, T. Hata*, *Tetrahedron Lett.*, 1985, **26**(*38*), 4621-4624

An = anisoyl
Ur = uridine

i) Me$_3$SiCl
ii) NH$_3$ - pyridine
60°, 3 h
iii) HCl, H$_2$O
20°, 24 h
→ UpU

81 %

Oxidation of diols by bis(tributyltin) oxide and bromine

R. Ravindran, T. R. Balasubramanian*, *Indian J. Chem., Sect. B*, 1986, **25B**(*11*), 1093-1094

R, R'= H, Ph

45 - 71 %
2 examples

56 - 87 %

Synthesis of silylbicycloalkenes

M. Grignon-Dubois, J. Dunogues, M. Ahra, *Recl. Trav. Chim. Pays-Bas*, 1988, **107**(*3*), 216-225

5 examples

88 %

85 %
exo/endo = 50/50

45 %   15 %

11. Bis(trimethylsilyl)acetamide (BSA)

CAS Registry Number	10416-59-8
CAS Name	Acetamide, N,O-bis(trimethylsilyl)-
Molecular Formula	$Me_3SiN=C(Me)OSiMe_3$
Molecular Weight	203.43
Boiling Point	71-73°C/35 mmHg
Melting Point	Not available.
Density	0.823 kg/m^3
Refractive Index	1.4170
Safety and Handling	Flammable liquid. Moisture sensitive. Irritant. Fp 11°C
Reactions	Silylating agent. Cyclocondensation. Review: *Synthesis*, 1985, 817.
Availability	Aldrich: p.
	Lancaster Synthesis: 98+%, p, bulk prices available.
	Sigma: p; sealed ampoules (1 ml ea.), ££.

Conversion of H-phosphonate mono- or diesters of nucleic acids into phosphate di- or triesters

E. de Vroom, M. L. Spierenburg, C. E. Dreef, G. A. van der Marel, J. H. van Boom*, *Recl. Trav. Chim. Pays-Bas*, 1987, **106**(2), 65-66

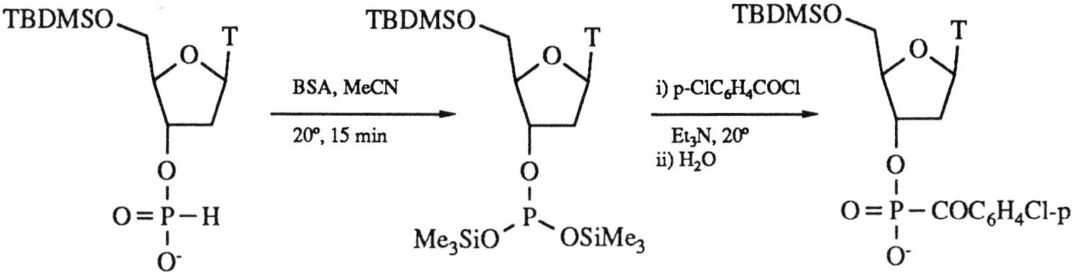

Preparation of cyclic amidines using BSA

K. Higashi, M. Sato*, M. Furukawa, *Chem. Pharm. Bull.*, 1986, **34**(*12*), 4927-4933

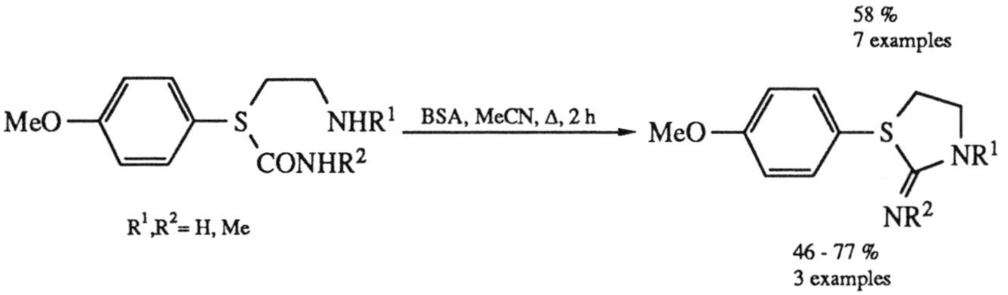

Regioselective reaction of acetoxy allyl phosphonates with nucleophiles catalyzed by palladium(0) complex

J. Zhu, X. Lu*, *Tetrahedron Lett.*, 1987, **28**(*17*), 1897-1900

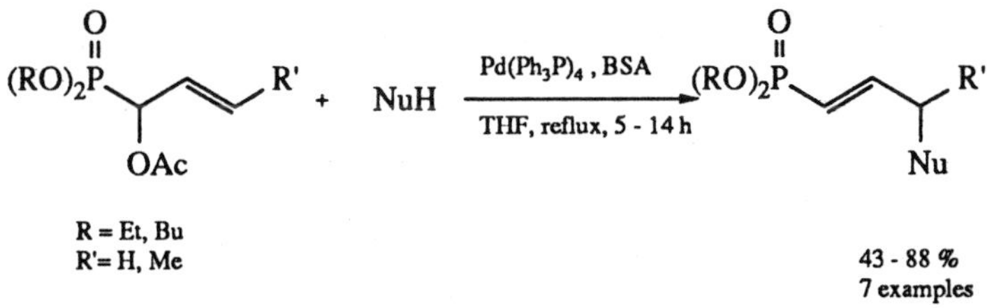

12. Bis(trimethylsilyl) peroxide

CAS Registry Number	5796-98-5
CAS Name	Silane, dioxy bis[trimethyl-
Molecular Formula	$(Me_3SiO)_2$
Molecular Weight	178.38
Boiling Point	$42°$ / 30 mm Hg
Melting Point	Not available.
Density	Not available.
Refractive Index	Not available.
Safety and Handling	Not available.
Reactions	Oxidation. Isomerization. Hydroxylation. A synthon for the hydroxyl cation.
Availability	Not commercially available.
Preparation	*J. Organomet. Chem.*, 1975, **99**(2), C31-C32.
Other Preparations	*J. Organomet. Chem.*, 1980, **201**(*1*), 197-211; *Zh. Strukt. Khim.*, 1981, **22**(*4*), 9-15; *Bull. Chem. Soc. Jpn.*, 1985, **58**(*3*), 844-9; *Synthesis*, 1986, (*8*), 633-5.

Isomerization of allylic alcohols

S. Matsubara, T. Okazoe, K. Oshima, K. Takai*, H. Nozaki, *Bull. Chem. Soc. Jpn.*, 1985, **58**(*3*), 844-849

$VO(acac)_2$, $(Me_3SiO)_2$

CH_2Cl_2 , r.t., 5 h

85 %
14 examples

Bis(trimethylsilyl) peroxide as a versatile reagent for stereoselective oxidation of phosphines to phosphine oxides

L. Wozniak, J. Kowalski, J. Chojnowski*, *Tetrahedron Lett.*, 1985, **26**(*40*), 4965-4968

$(Me_3SiO)_2$, $AlCl_3$

C_6H_6 , r.t., 1 h

95 % stereospecificity
94 % yield
5 examples

Electrophilic hydroxylation of organolithium compounds with bis(trimethylsilyl) peroxide

P. Molina*, A. Tarraga, M. J. Lidon, *Synthesis*, 1986, (*8*), 633-635

$$RX \xrightarrow[\text{ii) } (Me_3SiO)_2 \text{ , HCl} \atop \text{MeOH, r.t.}]{\text{i) BuLi}} ROSiMe_3 \xrightarrow[\text{15 min - 18 h}]{\text{HCl, MeOH, r.t.}} ROH$$

39 - 98 %
12 examples

13. 9-Borabicyclo[3.3.1]nonane (9-BBN)

CAS Registry Number   280-64-8, 21205-91-4 (dimer)

CAS Name   9-Borabicyclo[3.3.1]nonane

Molecular Formula

Molecular Weight   122.02, 244.04 (dimer)

Boiling Point   Not available.

Melting Point   150-152°C (dimer)

Density   0.894 kg/m^3 (THF soln)

Refractive Index   Not available.

Safety and Handling   Flammable liquid. Moisture sensitive. Dimer: Flammable solid.

Fp -22°C (hexanes), -17°C (THF)

Reactions   Hydroborating agent (dimer): *J. Am. Chem. Soc.*, 1974, **96**, 7765.
Reviews: G. W. Kabalka, *J. Organomet. Chem.*, 1987, **318**(1-3), 1-28; **337**(1-3), 169-194.

Availability   Aldrich: dimer, crystalline, 98%, £: 0.5M in hexanes, under N$_2$ in Sure/Seal™ bottles, p; 0.5M in THF under N$_2$ in Sure/Seal™ bottles and Kilo/Lab™ cylinders, p (cylinders require deposit).

Sigma: dimer, ££.

Preparation of (Z)-alkenes, ketones and alkynes by trialkyltin chloride induced intramolecular transfer reaction of lithium 1-alkynyltrialkylborates

K. K. Wang*, K.-H. Chu, *J. Org. Chem.*, 1984, **49**(26), 5175-5178

$$BuCH=CH_2 \xrightarrow[\substack{\text{ii) LiC} \equiv \text{CBu, Bu}_3\text{SnCl} \\ \text{reflux, 3 h}}]{\substack{\text{i) 9-BBN, THF, 2 h, r.t.} \\ \text{iii) AcOH, reflux, 1 h}}} \begin{array}{c} Bu \diagdown\diagup (CH_2)_5Me \\ \diagdown\diagup \\ 58\ \% \\ 14\ \text{examples} \end{array}$$

Stereoselective synthesis of (trimethylsilylmethyl)butadienes

C. Liu, K. K. Wang*, *J. Org. Chem.*, 1986, **51**(24), 4733-4734

R = H, alkyl
R'= alkyl

i) 9-BBN, THF r.t., 2 h

ii) R·C(O)·R' r.t., 1 h

KOH, H₂O₂, H₂O   58 - 88 %   7 examples

H₂SO₄   50 - 87 %   3 examples

One-pot transformation of carboxylic acids into aldehydes via acyloxyborabicyclononanes

J. S. Cha*, J. E. Kim, M. S. Yook, Y. S. Kim, *Tetrahedron Lett.*, 1987, **28**(49), 6231-6234

$$RCO_2H \xrightarrow{\text{9-BBN, THF, r.t.}} \left[RCO_2 - 9BBN \right] \xrightarrow[\substack{\text{-20°, 10 min} \\ \text{ii) 9-BBN, r.t., 1 h}}]{\text{i) t-BuLi, THF, pentane}} RCHO$$

R = alkyl, aryl   72 - 99 %   20 examples

14. Borane

CAS Registry Number   13283-31-3

CAS Name   Borane

Molecular Formula BH_3

Molecular Weight   13.8

Boiling Point   Not available.

Melting Point   Not available.

Density 0.898 kg/m^3 (THF soln.)

Refractive Index   Not available.

Safety and Handling   Flammable liquid. Moisture sensitive. THF complex may decompose in storage with liberation of hydrogen and bursting of bottle.

Fp -17°C

Reactions   Reduction. Hydroboration.

Availability   Aldrich: 1M in THF, under N_2 in Sure/Seal™ bottles and Kilo-Lab™ cylinders, p, (cylinders require deposit), stabilized with <0.005M $NaBH_4$. Also available as complexes of NH_3, Me_2S, phosphines and a wide variety of amines.

Lancaster Synthesis: Me_2S complex, 4% excess Me_2S, p, bulk prices available.

A stable catalyst for the enantioselective reduction of ketones

E. J. Corey, R. R. Bakshi, S. Shibata, C.-P. Chen, V. K. Singh, *J. Am. Chem. Soc.*, 1987, **109**(25), 7925-7926

$$R^1R^2CHOH$$

83.3 - 97.6 % ee
10 examples

R^1, R^2= alkyl, aryl

BH$_3$, THF
-10 to 32°, 2 - 25 min

Asymmetric reduction of oxime ethers

Y. Sakito*, Y. Yoneyoshi, G. Suzukamo*, *Tetrahedron Lett.*, 1988, **29**(2), 223-224

BH$_3$, THF, 20°, 12 h

R^1, R^2= alkyl, aryl
R^3= alkyl, aryl, silyl

40 - 82 % yield
79 - 92 % ee
11 examples

Use of acyloxyborane as an activating device for unsaturated carboxylic acids in their reactions with dienes

K. Furuta, Y. Miwa, K. Iwanaga, H. Yamamoto*, *J. Am. Chem. Soc.*, 1988, **110**(18), 6254-6255

+ BH$_3$.THF $\xrightarrow{CH_2Cl_2}$

-78°, 28 h

72 %
7 examples

15. Boron tribromide

CAS Registry Number 10294-33-4

CAS Name Borane, tribromo-

Molecular Formula BBr_3

Molecular Weight 250.54

Boiling Point 90^oC

Melting Point -46^oC

Density $2.560 \ kg/m^3$

Refractive Index 1.5312

Safety and Handling Corrosive. Moisture sensitive. Reacts violently with water. Very toxic by inhalation and swallowing. Causes severe burns.

Reactions Mild cleavage and brominating agent.

Availability Aldrich: 99.999%, £; 99.99%, p; 99+%, p; 1M in CH_2Cl_2, p; 1M in hexanes, p, all under N_2 in ampoules or Sure/Seal™ bottles. Also available as Me_2S complex.

Halogenative deoxygenation of ketones to vinyl bromides and/or *gem*-dibromides

E. Napolitano*, R. Fiaschi, E. Mastrorilli, *Synthesis*, 1986, (2), 122-125

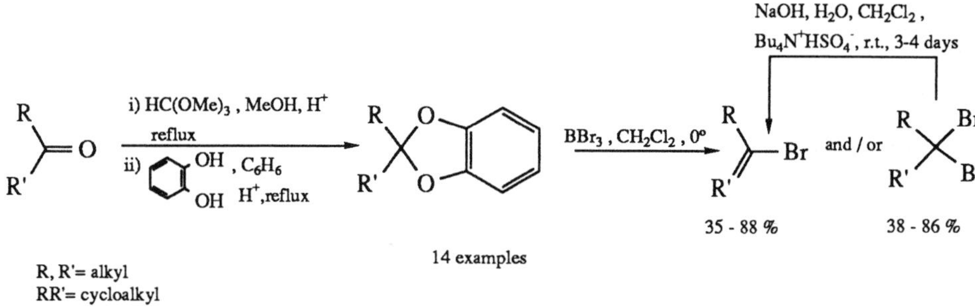

R, R'= alkyl
RR'= cycloalkyl

14 examples

35 - 88 %   38 - 86 %

Synthesis of unsymmetrical biphenyls and *m*-terphenyls

M. J. Sharp, W. Cheng, V. Snieckus*, *Tetrahedron Lett.*, 1987, **28**(*43*), 5093-5096, 5097-5098

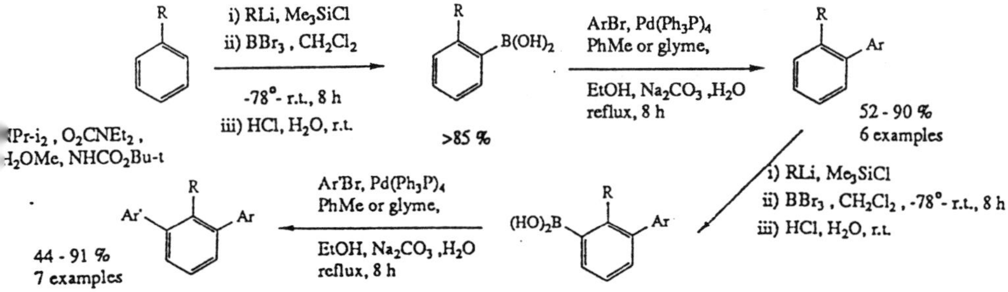

Direct conversion of silyl ethers into alkyl bromides with boron tribromide

S. Kim*, J. H. Park, *J. Org. Chem.*, 1988, **53**(*13*), 3111-3113

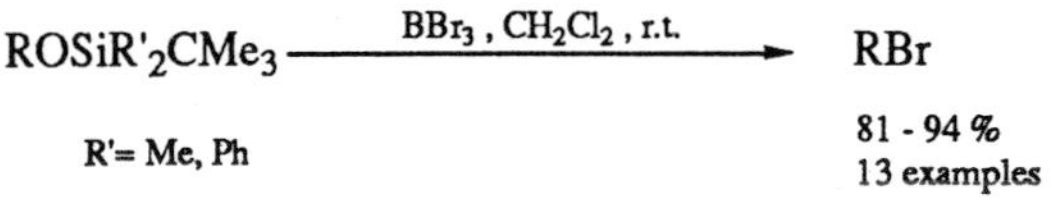

$$ROSiR'_2CMe_3 \xrightarrow{BBr_3 , CH_2Cl_2 , r.t.} RBr$$

R'= Me, Ph

81 - 94 %
13 examples

16. Boron trifluoride etherate

CAS Registry Number	109-63-7
CAS Name	Borane, trifluoro[1,1'-oxybis[ethane]]-, (T-4)-
Molecular Formula	$BF_3.OEt_2$
Molecular Weight	141.93, 67.81 (BF_3)
Boiling Point	126°C, -100°C (BF_3)
Melting Point	-58°C, -126.7°C (BF_3)
Density	1.154 kg/m^3, 2.99 kg/m^3 (BF_3)
Refractive Index	1.3480
Safety and Handling	Very toxic by inhalation. Causes severe burns. Reacts violently with water. Reacts with hot alkali or alkaline earth (not Mg) metals with incandescence. Fp 47°C
Reactions	Lewis acid catalyst. Cleavage of various ethers and protecting groups.
Availability	Aldrich: purified, redistilled, under N_2 in Sure/Seal™ bottles, p; in polycoated bottles, p. Also available as dibutyl or dimethyl etherate, $EtNH_2$, MeOH, Me_2S and Pr^{n}OH complexes. Lancaster Synthesis: p. Sigma: brown liquid, p: redistilled, light yellow liquid, p. Also available as MeOH complex.

Boron trifluoride promoted ring opening of oxiranes and oxetanes with fluorovinyllithiums. Synthesis of fluoro enols

T. Dubuffet, R. Sauvetre*, J.-F. No, *J. Organomet. Chem.*, 1988, **341**(*1-3*), 11-18

Bu⌒△O + F₂C=CFLi →[BF₃.Et₂O, -110°; / -80°, 15 min; / -100°, 20 min] Bu~CF₂ / OH F

88 %
17 examples

New method for the preparation of tertiary butyl ethers and esters

A. Armstrong, I. Brackenridge, R. F. W. Jackson*, J. M. Kirk, *Tetrahedron Lett.*, 1988, **29**(*20*), 2483-2486

Cl₃C / NH / OBu-t

→ ROH, BF₃.Et₂O, CH₂Cl₂ or cyclohexane, 16 -21 h → ROBu-t
58 - 91 %
7 examples

→ RCO₂H, BF₃.Et₂O, CH₂Cl₂ or cyclohexane, 16 h → RCO₂Bu-t
54 - 97 %
9 examples

Addition of aryllithium compounds to oxime ethers

K. E. Rodriques*, A. Basha, J. B. Summers, D. W. Brooks, *Tetrahedron Lett.*, 1988, **29**(*28*), 3455-3458

Me / NOCH₂Ph / H + PhLi →[BF₃.Et₂O, THF, N₂ / -78°, 30 min] Me / NHOCH₂Ph / Ph

44 %
8 examples

17. *N*-Bromosuccinimide (NBS)

CAS Registry Number   128-08-5

CAS Name   2,5-Pyrrolidinedione, 1-bromo-

Molecular Formula

Molecular Weight   177.99

Boiling Point   Not available.

Melting Point   177-181°C (dec.)/180-183°C

Density   Not available.

Refractive Index   Not available.

Safety and Handling   Moisture sensitive. Irritant. Causes burns.

Reactions   Selective brominating agent.
Reviews: *Chem. Rev.*, 1948, **43**, 271; 1970, **70**, 639; *Org. React.*, 1983, **29**, 1.

Availability   Aldrich: 99%, p.

Lancaster Synthesis: 99%, p, bulk prices available.

Sigma: crystalline, p.

Asymmetric halogenation of chiral imide enolates. Synthesis of enantiomerically pure α-amino acids

D. A. Evans*, J. A. Ellman, R. L. Dorow, *Tetrahedron Lett.*, 1987, **28**(*11*), 1123-1126

R = alkyl, aryl

78 - 96 % selectivity

95 - 100 %
99 % selectivity

67 - 86 %
78 - 95 % selectivity

One-pot synthesis of glycosyl bromides and acetates from benzyl glycosides with NBS

H. Hashimoto*, M. Kawa, Y. Saito, T. Date, S. Horito, J. Yoshimura, *Tetrahedron Lett.*, 1987, **28**(*30*), 3505-3508

87 %

9 examples

53 %

A new route to substituted 3-methoxycarbonyldihydropyrans

S. Hatakeyama, N. Ochi, H. Numata, S. Takano*, *J. Chem. Soc., Chem. Commun.*, 1988, (*17*), 1202-1204

R'=H, Me

83 - 94 %
5 examples

64 - 80 %
5 examples

34

18. *tert*-Butyl hydroperoxide (TBHP)

CAS Registry Number 75-91-2

CAS Name Hydroperoxide, 1,1-dimethylethyl

Molecular Formula Me_3COOH

Molecular Weight 90.12

Boiling Point 89°C (dec.)

Melting Point 6°C

Density 0.901 kg/m^3

Refractive Index 1.3980

Safety and Handling Flammable. Oxidizer. Harmful. Irritating to eyes and skin. Stable below 75°C. Liable to explode when distilled.

Fp 27°C (36°C)

Reactions Selective oxygenation reagent. Epoxidizer in anhydrous form.
Review: *Aldrichim. Acta*, 1979, **12**(*4*), 63.

Availability Aldrich: 90% (5% H$_2$O, 5% Bu^{t}OH), 70% (remainder H$_2$O), p; anhydrous, 3M in 2,2,4-trimethylpentane, p.

Lancaster Synthesis: 70% aq. soln, p. For azeotropic drying, see Lancaster Synthesis catalogue and *J. Org. Chem.*, 1983, **48**, 3607.

Sigma: 70% aq. soln, p.

Synthesis of optically pure γ-iodo allyl alcohols

Y. Kitano*, T. Matsumoto, T. Wakasa, S. Okamoto, T. Shimazaki, Y. Kobayashi, F. Sato*, K. Miyaji, K. Arai, *Tetrahedron Lett.*, 1987, 28(50), 6351-6354

t-BuOOH, Ti(OPr-i)$_4$,
D-(-)-DIPT, -20°, 42 h

(S)-
>99 % e.e.

>99 % e.e.

Improved synthesis of 2,5-cyclohexadien-1-ones via bis-allylic oxidation of 1,4-cyclohexadienes

A. G. Schultz*, A. G. Taveras, R. E. Harrington, *Tetrahedron Lett.*, 1988, 29(32), 3907-3910

t-BuOOH, PDC, CHCl$_3$, 25°

74 %
18 examples

Enantioselective synthesis of epoxynaphthoquinones via asymmetric Weitz-Scheffer epoxidation promoted by bovine serum albumin

S. Colonna, N. Gaggero, A. Manfredi, M. Spadoni, L. Casella, G. Carrea, P. Pasta, *Tetrahedron*, 1988, 44(16), 5169-5178

t-BuOOH, BSA
2 - 3 days

29 - 66 % yield
5 - 100 % ee
8 examples

BSA = bovine serum albumin
R = alkyl, Ph

19. Caesium fluoride

CAS Registry Number	13400-13-0
CAS Name	Cesium fluoride (CsF)
Molecular Formula	CsF
Molecular Weight	151.90
Boiling Point	1251°C
Melting Point	682°C
Density	4.115 kg/m^3
Refractive Index	1.478
Safety and Handling	Irritant. Hygroscopic.
Reactions	Horner-Emmons reaction. Fluoride ion-activated alkylation. Fluorination.
Availability	Aldrich: 99.99%, ££; 99.9%, £; 99%, p.
	Johnson Matthey: crystalline powder, Puratronic®, £; pieces, 99.99%, £; 99%, p.
	Sigma: crystalline, £.

Fluoride ion induced Horner-Emmons reaction of α-silylalkylphosphonic derivatives with carbonyl compounds

T. Kawashima*, T. Ishii, N. Inamoto*, *Bull. Chem. Soc. Jpn.*, 1987, **60**(*5*), 1831-1837

$$Ph_2C = O \;+\; (MeO)_2P(O)CHPhSiMe_3 \xrightarrow{\text{CsF, THF, reflux, 5 h}} PhCH = CPh_2$$

79 %
6 examples

Regiospecific and stereoselective alkylation of aldehydes with allyltrifluorosilane activated by fluoride ions

M. Kira*, M. Kobayashi, H. Sakurai*, *Tetrahedron Lett.*, 1987, **28**(*35*), 4081-4084

$$R^1CMe = CHCH_2SiF_3 \;+\; R^2CHO \xrightarrow[\text{6 - 36 h}]{\text{CsF, THF, r.t.- reflux,}} H_2C = CHCMeR^1CH(OH)R^2$$

88 - 96 %
12 examples

$R^1 = H, Me$
$R^2 = alkyl, aryl$

Fluoride-initiated reactions of α-halosilanes. Synthesis of stilbenes, epoxides and cyclopropanes

S. V. Kessar*, P. Singh, D. Venugopal, *Indian J. Chem., Sect. B,* 1987, **26**(*7*), 605-606

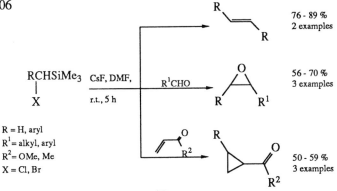

76 - 89 %
2 examples

56 - 70 %
3 examples

50 - 59 %
3 examples

$R = H, aryl$
$R^1 = alkyl, aryl$
$R^2 = OMe, Me$
$X = Cl, Br$

20. Camphorsulphonic acid (CSA)

CAS Registry Number 3144-16-9, 35963-20-3, 5872-08-2

CAS Name Bicyclo[2.2.1]heptane-1-methanesulfonic acid, 7,7-dimethyl-2-oxo-

Molecular Formula

Molecular Weight 232.30

Boiling Point Not available.

Melting Point 198°C (dec.) (1R)-(-), 206°C (dec.) (±)-, 194°C (dec.) (1S)-(+)

Density Not available.

Specific Rotation $[\alpha]_D^{20}$ -21° (c=2, H_2O) (1R)-(-), $[\alpha]_D^{20}$ +19.9° (c=2, H_2O) (1S)-(+)

Safety and Handling Corrosive. Hygroscopic. Irritant.

Reactions Resolving agent: (1S)-(+).

Availability Aldrich: (1R)-(-), 98%, p; monohydrate, 98%, p; (±), anhydrous, 98%, p; (1S)-(+), 99%, p.

Lancaster Synthesis: (±), 98%, p; (1S)-(+), monohydrate, 99%, p, bulk prices available.

Sigma: (1R)-(-), p; (1S)-(+), p.

Kinetic resolution of lactones with camphorsulphonic acid

K. Fuji*, M. Node, M. Murata, S. Terada, K. Hashimoto, *Tetrahedron Lett.*, 1986, **27**(*44*), 5381-5382

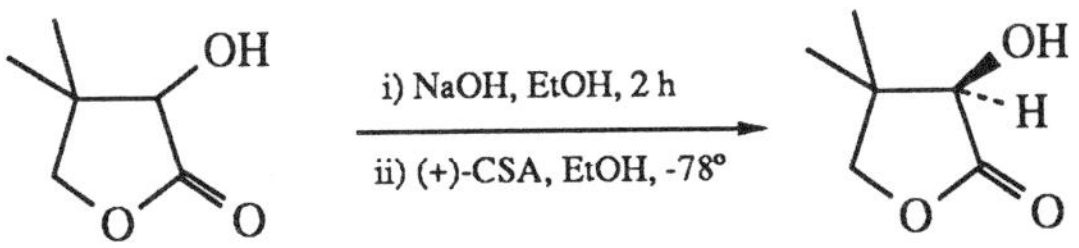
OH
i) NaOH, EtOH, 2 h
ii) (+)-CSA, EtOH, -78°
O
O
OH
H
O
O

(R)-(-)-
61.5 %
88.4 % ee

Nucleophile-promoted electrophilic cyclization of amino alkynes

L. E. Overman*, M. J. Sharp, *J. Am. Chem. Soc.*, 1988, **110**(2), 612-614

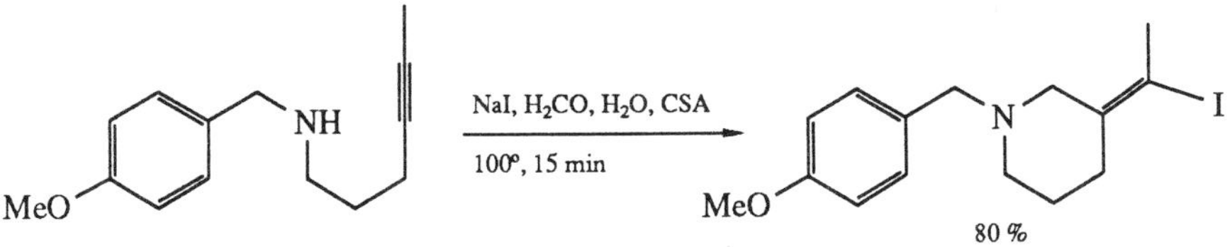
MeO
NH
NaI, H2CO, H2O, CSA
100°, 15 min
MeO
N
I

80 %
11 examples

A convenient, regioselective monoprotection of diols

M. Takasu, Y. Naruse, H. Yamamoto*, *Tetrahedron Lett.*, 1988, **29**(*16*), 1947-1950

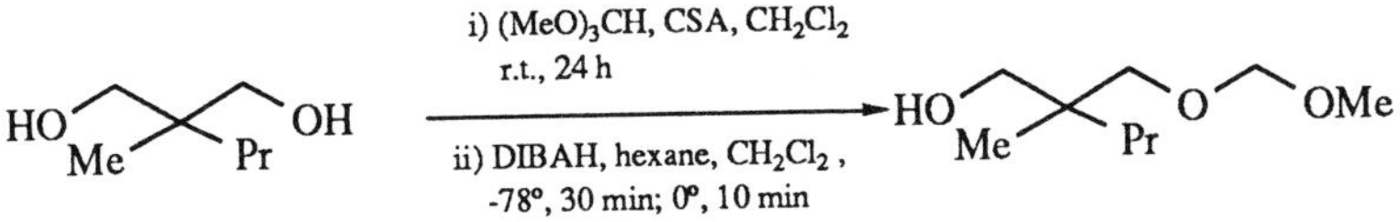
HO
Me
Pr
OH
i) (MeO)3CH, CSA, CH2Cl2
r.t., 24 h
ii) DIBAH, hexane, CH2Cl2,
-78°, 30 min; 0°, 10 min
HO
Me
Pr
O
OMe

92 %
14 examples

21. Carbon disulphide

CAS Registry Number	75-15-0
CAS Name	Carbon disulphide
Molecular Formula	CS_2
Molecular Weight	76.14
Boiling Point	$46^{o}C$
Melting Point	$-112^{o}C$
Density	1.266 kg/m^3
Refractive Index	1.6270
Safety and Handling	Extremely flammable. Very toxic by inhalation. Fp $-33^{o}C$
Reactions	Addition. Condensation. Insertion. Review: M. Yokoyama, T. Imamoto, *Synthesis*, 1984, (*10*), 797-824.
Availability	Aldrich, 99.9+%, HPLC grade (800 ml unit in Sure/Seal™ bottle), p; 99+%, spectrophotometric grade, p; ACS reagent, p.

Thiophenes by rearrangement of silyloxycyclopropane carboxylates

C. Bruckner, H.-U. Reissig*, *Angew. Chem., Int. Ed. Engl.*, 1985, **24**(7), 588-589

Reagents:
i) LDA, THF
ii) CS$_2$, THF, -78°, 1 h; 20°, 3 h
iii) MeI, 20°
iv) Et$_3$N.3HF, THF, 20°, 1 h

42 %
3 examples

Synthesis of alkyl dimethylhydrazono alkanedithioates

A. Oliva*, P. Delgado, *Synthesis*, 1986, (*10*), 865-866

R^1= Me, Ph
R^2= alkyl

i) BuLi, hexane, THF, -78°, 20 min
ii) CS$_2$, THF, -78°, 1 h; 20°, 4 h
iii) R^{2}X, THF, 20°, 17 h

75 - 90 %
8 examples

Reduction of nitro compounds to oximes

D. H. R. Barton*, I. Fernandez, C. S. Richard, S. Z. Zard, *Tetrahedron*, 1987, **43**(3), 551-558

CS$_2$, Et$_3$N, MeCN or CH$_2$Cl$_2$
0 - 20°, 0.5 - 84 h

R, R'= H, alkyl

29 - 85 %
11 examples

22. 1,1'-Carbonyldiimidazole (Carbodiimidazole)

CAS Registry Number   530-62-1

CAS Name   1*H*-Imidazole, 1,1-carbonylbis-

Molecular Formula

Molecular Weight   162.15

Boiling Point   Not available.

Melting Point   118-120°C

Density   Not available.

Refractive Index   Not available.

Safety and Handling   Harmful. Irritant. Moisture sensitive. Handle with exclusion of atmospheric moisture - hydrolyzed by water in a few seconds with evolution of CO_2.

Reactions   Prepared by reaction of imidazole in dry THF with phosgene in dry benzene: H. A. Staab, K. Wendel, *Ber.*, 1963, **96**(*12*), 3374. Peptide-coupling reagent. Reagent for carbonate and ester formation. Oxidizing agent. Dehydrating agent.

Availability   Aldrich: £.

Lancaster Synthesis: £, bulk prices available.

Sigma: £.

Esterification of carboxylic acids using carbonyldiimidazole and reactive halides

T. Kamijo*, H. Harada, K. Iizuka*, *Chem. Pharm. Bull.*, 1984, **32**(*12*), 5044-5047

$$RCO_2H \xrightarrow[\substack{\text{carbonyldiimidazole ,CHCl}_3 \text{ or MeCN} \\ \text{r.t. - reflux, 1 - 10 h}}]{\text{R'OH, allyl bromide or MeI}} RCO_2R'$$

R, R'= alkyl, aryl   79 - 95 %

A facile synthesis of thiazolidinones

M. d'Ischia, G. Prota*, R. C. Rotteveel, W. Westerhof, *Synth. Commun.*, 1987, **17**(*13*), 1577-1585

$$R_2C-\underset{\underset{SH\ \ NH_3^+Cl^-}{|\quad\ |}}{CH}\text{—}R^1 \xrightarrow[\substack{\text{ii)Na}_2\text{CO}_3\text{, H}_2\text{O, pH 9, r.t., 1 h} \\ \text{iii) HCl, H}_2\text{O}}]{\substack{\text{i) carbonyldiimidazole, THF} \\ \text{r.t., 15 h}}}$$

R = H, Me
R'= H, CO$_2$Et, CO$_2$Me

63 - 86 %
3 examples

Single-step conversion of primary amines to piperazinediones

C. G. Kruse*, J. J. Troost, P. Cohen-Fernandes, H. van der Linden, J. D. Van Loon, *Recl. Trav. Chim. Pays-Bas*, 1988, **107**(*4*), 303-309

$$R-N\overset{\displaystyle -CO_2H}{\underset{\displaystyle -CO_2H}{}} + R'NH_2 \xrightarrow[\text{THF or dioxane, reflux, 16 - 72 h}]{\text{carbonyldiimidazole}}$$

R = PhCH$_2$, Ph
R'= alkyl, aryl

63 - 98 %
25 examples

23. Cerium(IV) ammonium nitrate (CAN, Ceric ammonium nitrate)

CAS Registry Number	16774-21-3
CAS Name	Cerate(2-), hexakis(nitrato-O)-, diammonium, (OC-6-11)-
Molecular Formula	$(NH_4)_2Ce(NO_3)_6$
Molecular Weight	548.23
Boiling Point	Not available.
Melting Point	Not available.
Density	Not available.
Refractive Index	Not available.
Safety and Handling	Oxidizer. Irritant.
Reactions	On silica gel, mild oxidizing agent of hydroquinones, catechols, phenols to quinones: *Synthesis*, 1985, 641. Review: *Synthesis*, 1973, 347-354.
Availability	Aldrich: 99.99%, p; ACS Reagent, 99+%, p; volumetric standard, 0.05N in 4 wt % nitric acid, p. Also available on alumina and silica gel.
	Johnson Matthey: Crystalline powder, 99.5%, p.
	Sigma: p.

Oxidative bisdecarboxylation of α-alkoxymalonic acids with cerium(IV)

R. G. Salomon*, S. Roy, M. F. Salomon, *Tetrahedron Lett.*, 1988, **29**(7), 769-772

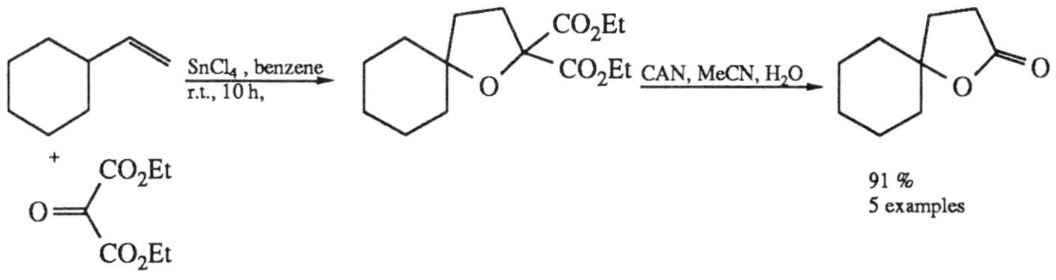

91 %
5 examples

Chemical and electrochemical synthesis of quinone imine *N*-oxides from indolinone arylimino nitroxides

A. Alberti, R. Andruzzi, L. Greci*, P. Stipa, G. Marrosu, A. Trazza, M. Poloni, *Tetrahedron*, 1988, **44**(5), 1503-1510

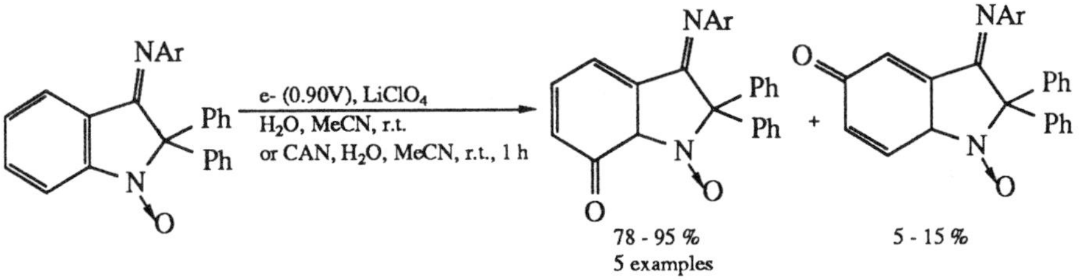

78 - 95 %
5 examples

5 - 15 %

Oxidative ring closure of allylsilanes with cerium(IV) ammonium nitrate

S. R. Wilson*, C. E. Augelli-Szafran, *Tetrahedron*, 1988, **44**(13), 3983-3995

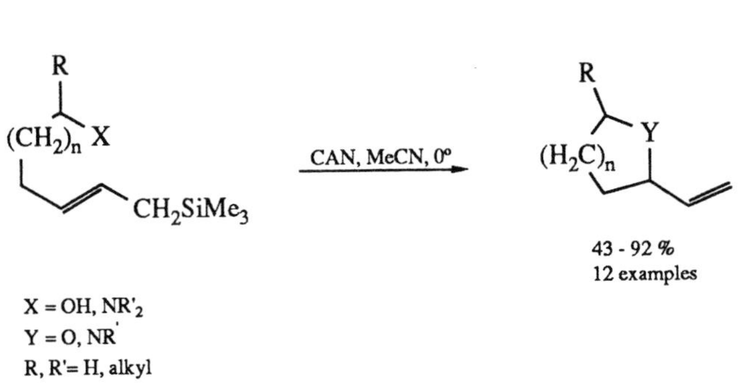

43 - 92 %
12 examples

X = OH, NR'$_2$
Y = O, NR'
R, R'= H, alkyl

46

24. Cerium(III) chloride (Cerous chloride)

CAS Registry Number	7790-86-5 (anhydrous), 18618-55-8 (heptahydrate)
CAS Name	Cerium chloride ($CeCl_3$)
Molecular Formula	$CeCl_3$
Molecular Weight	246.48, 372.59 (heptahydrate)
Boiling Point	1727°C
Melting Point	848°C (heptahydrate)
Density	3.970, 3.920 kg/m^3 (heptahydrate)
Refractive Index	Not available.
Safety and Handling	Hygroscopic.
Reactions	Mild selective reducing agent.
Availability	Aldrich: anhydrous, 99.9%, £; heptahydrate, 99.999%, ££; 99%, p.
	Johnson Matthey: hydrate, crystalline, 99.9%, p; hydrate, crystalline, 99.99%, p.
	Sigma: heptahydrate, crystalline, p.

Facile reduction of organic halides and phosphine oxides with lithium aluminium hydride and cerium trichloride

T. Imamoto*, T. Takeyama, T. Kusumoto, *Chem. Lett.*, 1985, (*10*), 1491

$$Me(CH_2)_{11}F \xrightarrow[\text{THF, reflux, 3 h}]{\text{LiAlH}_4 \text{, CeCl}_3} Me(CH_2)_{10}Me$$

90 %
16 examples

Reaction of α-halo ketones with carbonyl compounds promoted by cerium halides

S. Fukuzawa*, T. Tsuruta, T. Fujinami, S. Sakai, *J. Chem. Soc., Perkin Trans. 1*, 1987, (*7*), 1473-1477

i) CeCl$_3$ - NaI or CeCl$_3$ - SnCl$_2$
THF, r.t., 2 h
ii) HCl, H$_2$O

R, R^2= alkyl, aryl
R^1, R^3= H, alkyl

30 - 95 %
28 examples

Cerium-mediated conversion of esters to allylsilanes

B. A. Narayanan, W. H. Bunnelle*, *Tetrahedron Lett.*, 1987, **28**(*50*), 6261-6264

i) CeCl$_3$ / Me$_3$SiCH$_2$MgCl , -70°, 1 h
ii) -70°, 2 h; -70° to r.t., overnight
iii) silica gel, CH$_2$Cl$_2$, 2 - 3 h

R = alkyl, aryl, allyl
R'= Me, Et

77 - 95 %
6 examples

25. Cetyltrimethylammonium permanganate

CAS Registry Number   73257-07-5

CAS Name   1-Hexadecanaminium, *N,N,N*-trimethyl-, salt with permanganic acid (HMnO₄)(1:1)

Molecular Formula $C_{16}H_{33}Me_3N.MnO_4$

Molecular Weight   Not available.

Boiling Point   Not available.

Melting Point   Not available.

Density   Not available.

Refractive Index   Not available.

Safety and Handling   Not available.

Reactions   Oxidation. Hydroxylation. Oxidative cleavage.

Availability   Not commercially available.

Preparation   *Synthesis*, 1984, (*5*), 431-433.

Other Preparations   *Mikrochim. Acta*, 1979, **2**(*5-6*), 373-381; *J. Org. Chem.*, 1984, **49**(*23*), 4509-4516.

Oxidative cyclization of hydroxy alkenes with cetyltrimethylammonium permanganate. Synthesis of γ- and δ-lactones

R. Rathore, P. S. Vankar, S. Chandrasekaran*, *Tetrahedron Lett.*, 1986, **27**(*34*), 4079-4082

HO–[structure] $\xrightarrow{\text{Me(CH}_2)_{15}\text{Me}_3\text{N}^+\text{MnO}_4^-, \text{CH}_2\text{Cl}_2}_{\text{r.t., 3.5 h}}$ [lactone structure]

70 %
15 examples

Cetyltrimethylammonium permanganate: a useful reagent for selective oxidative cleavage of aryl alkenes

R. Rathore, S. Chandrasekaran*, *J. Chem. Res.*, 1986, (*12*), 458-459

Ph–CH=CH–Ph $\longrightarrow$ PhCHO 92 %

$\xrightarrow{\text{Me(CH}_2)_{15}\text{Me}_3\text{N}^+\text{MnO}_4^-, \text{CH}_2\text{Cl}_2}_{\text{r.t., 2 h}}$ 11 examples

Ph$_2$C=CH$_2$ $\longrightarrow$ Ph$_2$CO 94 %

A mild and selective method for the conversion of nitroalkanes to carbonyl compounds

P.S. Vankar, R. Rathore, S. Chandrasekaran*, *Synth. Commun.*, 1987, **17**(2), 195-201

$\underset{R \quad R'}{\overset{NO_2}{\diagdown}}$ $\xrightarrow{\text{Me(CH}_2)_{15}\text{Me}_3\text{N}^+\text{MnO}_4^-, \text{CH}_2\text{Cl}_2}_{\text{r.t., 3 - 5.5 h}}$ $\underset{R \quad R'}{\overset{O}{\diagdown}}$

R, R'= H, alkyl, aryl

57 - 89%
13 examples

50

26. 3-Chloroperoxybenzoic acid (mCPBA)

CAS Registry Number 937-14-4

CAS Name Benzenecarboperoxoic acid, 3-chloro-

Molecular Formula $ClC_6H_4CO_3H$

Molecular Weight 172.57

Boiling Point Not available.

Melting Point 92-94°C (dec.)

Density Not available.

Refractive Index Not available.

Safety and Handling Irritant. Oxidizer.

Reactions Versatile stable oxidizing agent.

Availability Aldrich: tech. grade, 50-55%, p. Contains 7-10% mCPBA, remainder H_2O.

Lancaster Synthesis: 50-55%; p. Contains 10% mCPBA, remainder H_2O.

Sigma: practical grade approx. 85%, crystalline, p.

Cleavage of dithioacetals to the corresponding carbonyl compounds

J. Cossy, *Synthesis*, 1987, (*12*), 1113-1115

$$R^1 \diagdown_{R^2}\diagup^{SR^3}_{SR^3} \xrightarrow{\text{mCPBA, } F_3CCO_2H, CH_2Cl_2, 0°} R^1 \diagdown_{\substack{\parallel \\ O}}\diagup R^2$$

R^1, R^2= H, alkyl
R^3= Ph, alkyl

10 - 80 %
9 examples

Oxidative conversion of β-hydroxy selenides to epoxides and ketones with mCPBA

S. Uemura*, K. Ohe, N. Sugita, *J. Chem. Soc., Chem. Commun.*, 1988, (2), 111-112

Me(CH$_2$)$_{12}$ $\diagdown$ SePh, Et, OH $\longrightarrow$ Me(CH$_2$)$_{12}$ $\diagdown$ Et $\triangle$ O   96 %

SePh, OH $\xrightarrow{\text{mCPBA, MeOH, THF r.t., 0.5 h}}$ O   74 %

8 examples

Preparation of α,β-unsaturated carbonyl compounds by peracid oxidation of isoxazolines

A. Padwa*, U. Chiacchio, D. N. Kline, J. Perumattan, *J. Org. Chem.*, 1988, 53(*10*), 2238-2245

$$R^1\diagdown_{R^2}\diagup^{N-R}_{O}\diagdown_{R^3} \xrightarrow{\text{mCPBA, } CH_2Cl_2, 0°} \substack{R^1 \\ H}\diagup\diagdown\substack{R^2 \\ COR^3}$$

R^1, R^3= alkyl, alkynyl, CO$_2$Me
R^2= electron withdrawing group

65 - 98 %
>10 examples

27. N-Chlorosuccinimide (NCS)

CAS Registry Number	128-09-6
CAS Name	2,5-Pyrrolinedione, 1-chloro-
Molecular Formula	
Molecular Weight	133.53
Boiling Point	Not available.
Melting Point	150-152°C
Density	1.65 kg/m³
Refractive Index	Not available.
Safety and Handling	Corrosive. Moisture sensitive. Violent or explosive reaction with aliphatic alcohols.
Reactions	Source of positive chlorine for oxidation and chlorination. Conversion of 1° and 2° alcohols to carbonyls, allylic and benzylic alcohols to halides.
Availability	Aldrich: 98+%, p.
	Lancaster Synthesis: 98+%, p, bulk prices available.
	Sigma: 95-98%, crystalline, p.

Ring cleavage of (α-phenylthiobenzyl)cycloalkanols

M. Yasumura*, K. Takaki, T. Tamura, K. Negoro, *Bull. Chem. Soc. Jpn.*, 1986, 59(*1*), 317-318

i) NCS, CH$_2$Cl$_2$, -40°, 3 h
ii) Et$_3$N, CH$_2$Cl$_2$, r.t., 3 h

n = 3 - 5
R = H, Me

10 - 76 %
6 examples

Preparation of α-acyloxy and α-halo sulphides via desilylative rearrangement of sulphides and sulphoxides

H. Ishibashi*, H. Nakatani, K. Maruyama, K. Minami, M. Ikeda, *J. Chem. Soc., Chem. Commun.*, 1987, (*19*), 1443-1445

NCS, F$_3$CCO$_2$H
or NBS, F$_3$CSO$_3$H
CCl$_4$, 0°

R = alkyl, aryl

X = Cl, Br
43 - 98 %
10 examples

Preparation of 1-chloroalkyl p-tolyl sulphoxides in high optical yields

T. Satoh, T. Oohara, Y. Ueda, K. Yamakawa*, *Tetrahedron Lett.*, 1988, 29(*3*), 313-316

NCS, K$_2$CO$_3$
CH$_2$Cl$_2$, r.t., 20 - 43 h

R = H, Me
R^1= H, Me, CH$_2$Ph

88 - 94 %
87 - 94 % ee
4 examples

28. Chlorosulphonyl isocyanate (*N*-Carbonylsulphamyl chloride)

CAS Registry Number	1189-71-5
CAS Name	Sulfuryl chloride isocyanate
Molecular Formula	ClSO$_2$NCO
Molecular Weight	141.53
Boiling Point	107°C
Melting Point	-44°C
Density	1.626 kg/m^3
Refractive Index	1.4467
Safety and Handling	Corrosive. Lachrymatory.
Reactions	Most chemically reactive isocyanate known. Forms β-lactams by cycloaddition to unactivated unsaturated hydrocarbons. Reviews: D. N. Dhar, K. S. K. Murthy, *Synthesis*, 1986, (*6*), 437-449; A. Kamal, P. B. Sattur, *Heterocycles*, 1987, **26**(*4*), 1051-1076.
Availability	Aldrich: 98%, p. Sigma: p.

Synthesis of 1,3-dioxolan-2-ones from epoxides using chlorosulphonyl isocyanate

K. S. K. Murthy, D. N. Dhar*, *Synth. Commun.*, 1984, **14**(*7*), 687-695

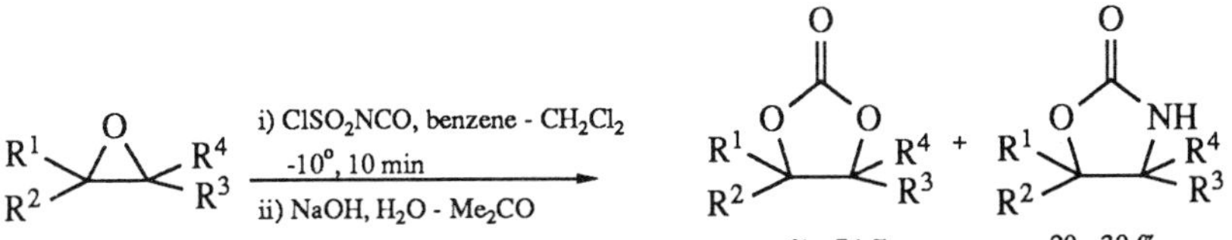

i) ClSO$_2$NCO, benzene - CH$_2$Cl$_2$
-10°, 10 min
ii) NaOH, H$_2$O - Me$_2$CO

R^1- R^4= H, aryl, alkyl

61 - 74 %
5 examples

20 - 30 %

Synthesis of primary 2-alkynamides

P. C. B. Page*, S. Rosenthal, R. V. Williams, *Synthesis*, 1988, (*8*), 621-623

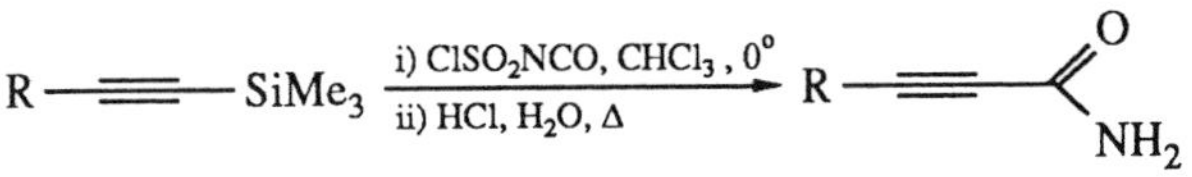

i) ClSO$_2$NCO, CHCl$_3$, 0°
ii) HCl, H$_2$O, Δ

R = alkyl, Ph

54 - 71 %
4 examples

Efficient synthesis of cyclic enamides and 2*H*-pyrroles

S. P. Joseph, D. N. Dhar*, *Tetrahedron*, 1988, **44**(*16*), 5209-5214

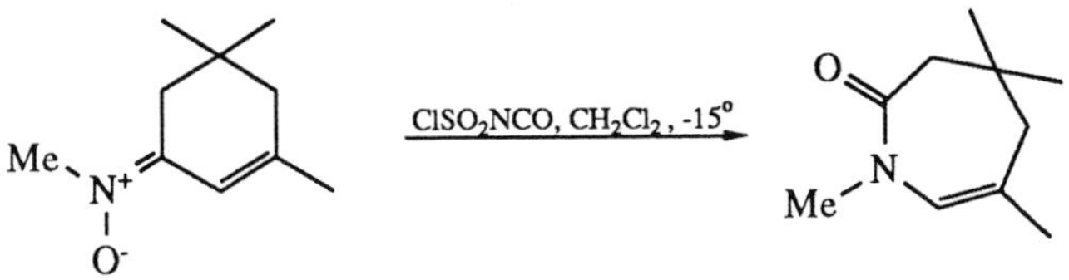

ClSO$_2$NCO, CH$_2$Cl$_2$, -15°

42 %
13 examples

56

29. Chlorotrimethylsilane

CAS Registry Number	75-77-4
CAS Name	Silane, chlorotrimethyl-
Molecular Formula	Me_3SiCl
Molecular Weight	108.64
Boiling Point	57°C
Melting Point	-40°C
Density	0.856 kg/m^3
Refractive Index	1.3870
Safety and Handling	Flammable liquid. Corrosive. Reacts violently with water. Irritating to eyes and respiratory system. Fp -18°C
Reactions	Reagent for preparation of volatile TMS ethers for GC analysis. Silylating agent. Reviews: R. Muller, *Z. Chem.*, 1985, **25**(*9*), 309-318; *Angew. Chem.*, 1965, **77**, 417; *Synthesis*, 1985, 817.
Availability	Aldrich: 98%, p. Lancaster Synthesis: 98+%, p, bulk prices available. Sigma: p.

Synthesis of alkyl iodides by hydroiodination of alkenes

S. Irifune, T. Kibayashi, Y. Ishii*, M. Ogawa, *Synthesis*, 1988, (5), 366-369

$$R^1R^3C=CR^2R^4 \xrightarrow[\text{MeCN, r.t., 1 - 3 h}]{\text{Me}_3\text{SiCl, NaI, H}_2\text{O}} R^1R^3C(I)-C(H)R^2R^4$$

R¹ - R³= H, alkyl
R⁴= H, CN, COMe

65 - 69 %
10 examples

Chlorination of ketones by trimethylchlorosilane and dimethyl sulphoxide with bromide ion catalysis

R. B. Fraser*, F. Kong, *Synth. Commun.*, 1988, **18**(*10*), 1071-1077

$$RCH_2COR' \xrightarrow[\text{THF or MeCN, r.t.}]{\text{Me}_3\text{SiCl, DMSO, R}^2{}_4\text{N}^+\text{Br}^-} RClCHCOR'$$

R, R'= alkyl, aryl
R²= alkyl

29 - 98 %
8 examples

Mannich reactions of nucleophilic aromatic compounds involving aminals and α-amino ethers activated by chlorosilane derivatives

H. Heaney*, G. Papageorgiou, R. F. Wilkins, *J. Chem. Soc., Chem. Commun.*, 1988, (*17*), 1161-1163

X = N, O
R, R¹= Me, H, C₄H₄
R² = allyl, cycloalkyl

15 - 86 %
21 examples

30. Chromium(II) chloride

CAS Registry Number   10049-05-5

CAS Name   Chromium chloride ($CrCl_2$)

Molecular Formula $CrCl_2$

Molecular Weight   122.90

Boiling Point   Not available.

Melting Point   824°C

Density   2.900 kg/m^3

Refractive Index   Not available.

Safety and Handling   Moisture sensitive. Irritant.

Reactions   Reduction.

Availability   Aldrich: anhydrous, tech. grade, ££. May contain 1-2% insoluble matter.

Johnson Matthey: powder, 98%, ££.

Chromium(II) mediated synthesis of homoallylic alcohols

P. G. M. Wutts*, G. R. Callen, *Synth. Commun.*, 1986, **16**(*14*), 1833-1837

$$RCHO \xrightarrow[\text{THF, 1 - 1.5 h}]{\text{BrCH}_2\text{CH=CHMe, CrCl}_2}$$

R = alkyl, aryl

59 - 99 %
threo: erythro 1.4 - 15.9:1
11 examples

Preparation of *ortho*-hydroxybenzyl ketones and benzofurans from acyloxybenzyl bromides via their chromium complexes

B. Ledoussal, A. Gorgues*, A. Le Coq, *Tetrahedron*, 1987, **43**(*24*), 5841-5852

i) CrCl$_2$, THF, reflux, 5 h
ii) NH$_4$Cl, H$_2$O

20 - 85 %

AcOH, HCl, Δ

i) CrCl$_2$, BF$_3$.OEt$_2$, THF
reflux, 5 h
ii) NH$_4$Cl, H$_2$O

15 - 96 %
13 examples

= alkyl, aryl
= H, Cl, OAc, OBz

Reduction of acrolein dialkyl acetals with chromium(II) chloride

K. Takai*, K. Nitta, K. Utimoto, *Tetrahedron Lett.*, 1988, **29**(*41*), 5263-5266

+ PhCHO $\xrightarrow[\text{THF, -30°, 3 h}]{\text{CrCl}_2, \text{Me}_3\text{SiI}}$

99 %, 88:12
13 examples

31. Copper(I) cyanide

CAS Registry Number	544-92-3
CAS Name	Copper cyanide (CuCN)
Molecular Formula	CuCN
Molecular Weight	89.56
Boiling Point	Not available.
Melting Point	$473^{o}C$ (in N_2)
Density	2.920 kg/m^3
Refractive Index	Not available.
Safety and Handling	Highly toxic. Irritant.
Reactions	Reagents for synthesis of organocuprates.
Availability	Aldrich: 99%, p.
	Johnson Matthey: crystalline, p.

Synthesis of enoates by copper(I)-catalyzed allylation of zinc esters

H. Ochiai, Y. Tamaru, K. Tsukaki, Z. Yoshida*, *J. Org. Chem.*, 1987, **52**(*19*), 4418-4420

$$IZn \overset{}{\frown} (CH_2)_n \overset{.CO_2Et}{} + R^1 \overset{R^2}{\underset{}{\frown}} X \xrightarrow[60°, 1 \text{ h or r.t., overnight}]{\text{CuCN, THF, DMA}} R^1 \overset{R^2}{\underset{}{\diagup\diagdown}} (CH_2)_{n+1}CO_2Et$$

$$+ R^1 \overset{}{\underset{R^2}{\diagup}} (CH_2)_{n+1}CO_2Et$$

n = 1, 3

$R^1, R^2 =$ H, alkyl, aryl
X = Ts, Cl, Br

50 - 99 %
28 : 72 - 0:100 mixtures

Synthesis of sodium organocuprates

S. H. Bertz*, C. P. Gibson, G. Dabbagh, *Organometallics*, 1988, **7**(*1*), 227-232

$$RNa \xrightarrow[\text{Ar, -50°, 12 - 15 min}]{\text{CuCN or CuBr. SMe}_2, \text{THF}} R_2CuNa$$

R = Bu, pentyl, PHCH$_2$, Ph

In-situ cuprate formation via transmetallation between vinylstannanes and higher order cyanocuprates. Application to prostaglandin synthesis

J. R. Behling, K. A. Babiak, J. S. Ng, A. L. Campbell*, R. Moretti, M. Koerner, B. H. Lipshutz*, *J. Am. Chem. Soc.*, 1988, **110**(*8*), 2641-2643

CuCN

i) MeLi, THF, Et$_2$O, 0°

ii) Bu$_3$Sn $\overset{Me}{\underset{OSiMe_3}{\diagdown\diagup}}$, THF, r.t., 1.5 h

iii) $\overset{O}{\diagdown}$ (CH$_2$)$_6$CO$_2$Me , THF; -64°, -35°, 3 min
Et$_3$SiO

iv) NH$_4$Cl, NH$_4$OH, H$_2$O

Misoprostanol
91 %
10 examples

32. 18-Crown-6

CAS Registry Number	17455-13-9
CAS Name	1,4,7,10,13,16-Hexaoxacyclooctadecane
Molecular Formula	

Molecular Weight	264.32
Boiling Point	Not available.
Melting Point	37-39°C (from MeCN)
Density	Not available.
Refractive Index	Not available.
Safety and Handling	Irritant. Moisture sensitive.
Reactions	Complexing agent to solubilize alkali metal ions in non-polar solvents. Reviews: *Synthesis*, 1976, 168; *J. Heterocycl. Chem.*, 1982, **19**, 3.
Availability	Aldrich: 99.5+%, Gold Label, £££; 99%, £.
	Lancaster Synthesis: 98+%, £, bulk prices available.
	Sigma: £.

Selective fluorination of methanols with methanesulphonyl fluoride, caesium fluoride and 18-crown-6

K. Makino, H. Yoshioka*, *J. Fluorine Chem.*, 1987, **35**(*4*), 677-683

$$ArCH_2OH \xrightarrow[\text{THF, reflux, 4 - 15 h}]{\text{CsF, MsF, 18-crown-6}} ArCH_2F$$

23 - 80 %
9 examples

Alkylation and acylation of trinitrobenzene with silyl enol ethers

G. A. Artamkina*, S. V. Kovalenko, I. P. Beletskaya, O. A. Reutov, *J. Organomet. Chem.*, 1987, **329**(*2*), 139-150

KF, 18-crown-6, THF
r.t., 3 h

K⁺.18-crown-6

93 %

NBS, THF
30 min

81 %
9 examples

Convenient route to enolate anions via a novel reaction between β-propiolactones and a solution of potassium containing 18-crown-6

Z. Jedlinski*, M. Kowalczuk, A. Misiolek, *J. Chem. Soc., Chem. Commun.*, 1988, (*18*), 1261-1262

K, 18-crown-6
THF, -20°

MeI

$MeCH_2 - \overset{O}{\overset{\|}{C}}O \cdot \overset{R}{\overset{|}{C}}HMe$

R = H, Me

65 - 70 %
2 examples

33. Cyanogen bromide (Bromine cyanide)

CAS Registry Number	506-68-3
CAS Name	Cyanogen bromide
Molecular Formula	BrCN
Molecular Weight	105.93
Boiling Point	61-62°C
Melting Point	49-51°C
Density	1.039 (MeCN soln)
Refractive Index	Not available.
Safety and Handling	Highly toxic. Irritant. Fp 5°C
Reactions	Cyanation. Bromination. Activating agent for insoluble supports for affinity absorption. *Anal. Biochem.*, 1974, **60**, 149.
Availability	Aldrich: 97%, p; 5M in MeCN, under N_2 in Sure/Seal™ bottles, p. Sigma: p. Also available on Sepharose support as affinity chromatography media.

Asymmetric synthesis of α-amino acids

S. K. Phadtare, S. K. Kamat, G. T. Panse*, *Indian J. Chem., Sect. B*, 1985, **24**(*8*), 811-814

HO + NH$_2$ $-$ C$\overset{\text{Me}}{\underset{\text{H}}{|}}$ $-$ Ph

R = alkyl, aryl

i) benzene, reflux, 1 - 6 h
ii) BrCN, Et$_2$O, 0°, 2 h;
   r.t., overnight
iii) Et$_3$N, r.t.

R $-$ C$\overset{\text{H}}{\underset{\text{CN}}{|}}$ $-$ N = C $-$ Ph (Me)

i) 6N HCl, reflux, 2 h
ii) Dowex 50 column,
    H$_2$O eluent

R $-$ C$\overset{\text{H}}{\underset{\text{CO}_2\text{H}}{|}}$ $-$ NH$_2$

20 - 65 % from Schiff's base
< 52 % optical purity
8 examples

Bromination of β-amino enones with cyanogen bromide

A. Alberola*, C. Andres, A. Gonzalez Ortega, R. Pedrosa, M. Vicente, *Synth. Commun.*, 1986, **16**(*10*), 1161-1165

BrCN, MeOH, 0°, 8 h

R^1= alkyl
R^2= H, Ph

50 - 96 %
6 examples

Cyanogen bromide-dimethylaminopyridine: a convenient source of positive cyanide for the synthesis of cyanoimidazoles

J. P. Whitten*, J. R. McCarthy, D. P. Matthews, *Synthesis*, 1988, (*6*), 470-472

DMF, 40°, 16 h

R^1= H, Me, Ph, Cl
R^2= alkyl

43 - 90 %
8 examples

34. Cyanotrimethylsilane

CAS Registry Number   7677-24-9

CAS Name   Silanecarbonitrile, trimethyl-

Molecular Formula Me_3SiCN

Molecular Weight   99.21

Boiling Point   118-119°C

Melting Point   11-12°C

Density   0.744 kg/m^3

Refractive Index   1.3924

Safety and Handling   Highly toxic. Flammable liquid.
Fp 1°C

Reactions   Cyanation.

Availability   Aldrich: 98%, ampoules, ££; 98%, under N_2 in Sure/Pac™ metal cylinders, £.

Asymmetric hydrocyanation of aldehydes with cyanotrimethylsilane promoted by chiral titanium reagent

K. Narasaka, T. Yamada, H. Minamikawa, *Chem. Lett.*, 1987, (*10*), 2073-2076

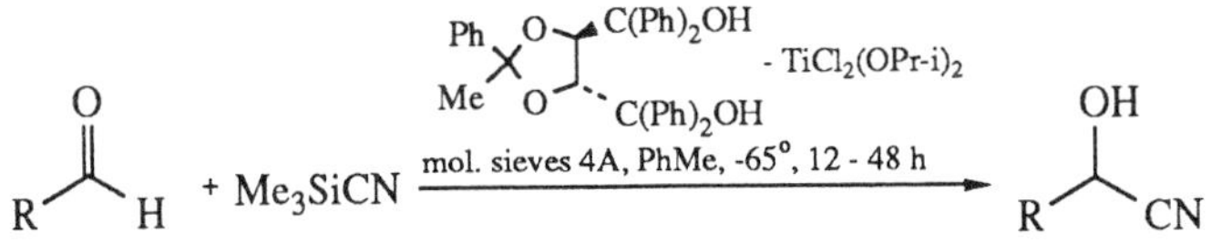

R = alkyl, aryl, cyclohexyl

66 - 89 %
68 - 90 % ee
5 examples

Preparation and cyclization of bis(arylamino)propenenitriles

M. Takahashi*, H. Miyahara, N. Yoshida, *Heterocycles*, 1988, 27(*1*), 155-171

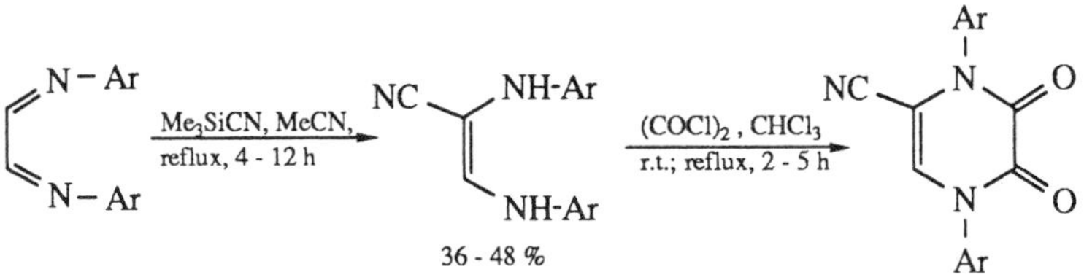

36 - 48 %

28 - 65 %
4 examples

Regioselective one-pot synthesis of cyanodeoxy sugars

S. N.-ul-H. Kazmi, Z. Ahmed, A. Q. Khan, A. Malik*, *Synth. Commun.*, 1988, 18(2), 151-156

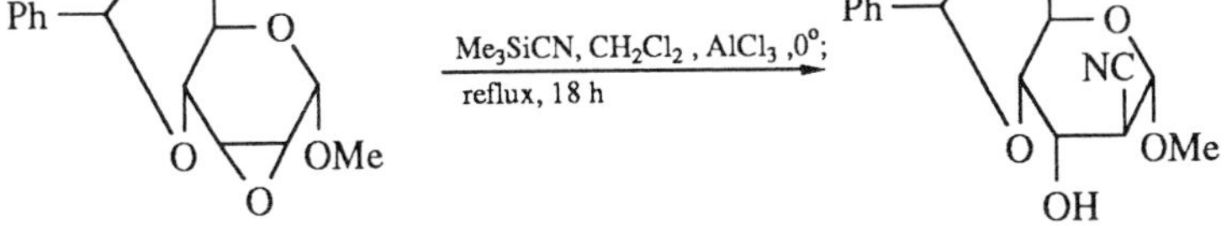

47 %
4 examples

35. 1,4-Diazabicyclo[2.2.2]octane (DABCO, Triethylenediamine, TED)

CAS Registry Number 280-57-9

CAS Name 1,4-Diazabicyclo[2.2.2]octane

Molecular Formula

Molecular Weight 112.18

Boiling Point Not available.

Melting Point 158-160°C

Density Not available.

Refractive Index Not available.

Safety and Handling Corrosive. Hygroscopic. Irritant.

Reactions Forms crystalline complexes with organolithiums. Cleaves β-keto esters directly to ketones. Catalyst for condensation of acrylates with aldehydes. Useful additive in organometallic reactions.

Availability Aldrich: 97%, p.

Lancaster Synthesis: 97+%, p, bulk prices available.

Sigma: p.

Mikanecic acid total synthesis

H. M. R. Hoffmann*, J. Rabe, *Helv. Chim. Acta*, 1984, **67**(2), 413-415

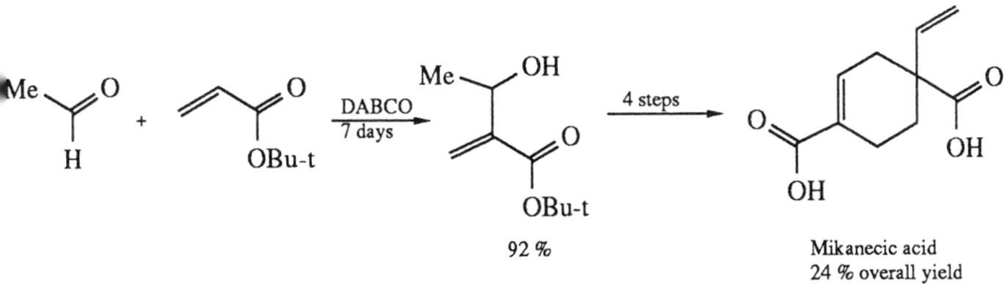

Me
O
H
+
O
OBu-t
DABCO
7 days
Me
OH
O
OBu-t
92 %
4 steps
O
OH
O
OH
Mikanecic acid
24 % overall yield

Stereoselective generation and facile dimerization of hydroxymethylenealkanoic esters

W. Poly, D. Schomburg, H. M. R. Hoffmann*, *J. Org. Chem.*, 1988, **53**(*16*), 3701-3710

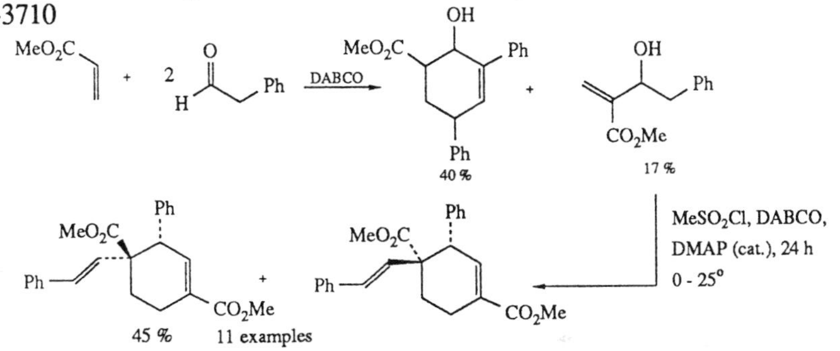

MeO2C
+
2
O
H
Ph
DABCO
OH
MeO2C
Ph
Ph
40 %
+
OH
Ph
CO2Me
17 %
MeSO2Cl, DABCO,
DMAP (cat.), 24 h
0 - 25°
Ph
MeO2C
Ph
CO2Me
45 %
+
11 examples
Ph
MeO2C
Ph
CO2Me

Stereoselective synthesis of α-methylene-β-hydroxy-γ-alkoxy esters and ketones

S. E. Drewes, T. Manickum, G. H. P. Roos*, *Synth. Commun.*, 1988, **18**(*10*), 1065-1070

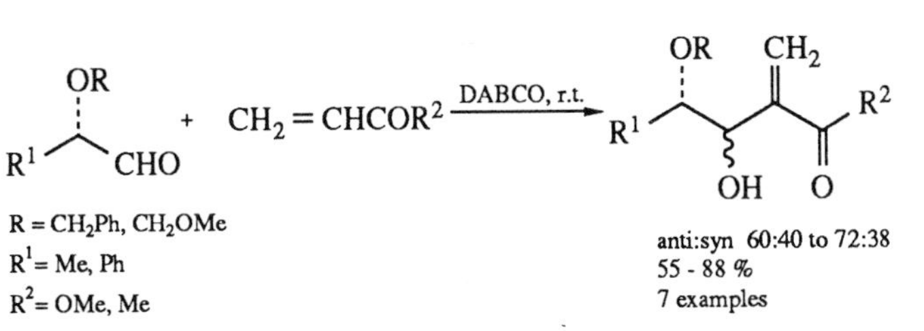

OR
R1
CHO
+
CH2 = CHCOR2
DABCO, r.t.
OR
CH2
R1
R2
OH
O
R = CH2Ph, CH2OMe
R1= Me, Ph
R2= OMe, Me
anti:syn 60:40 to 72:38
55 - 88 %
7 examples

36. 1,8-Diazabicyclo[5.4.0]undec-7-ene (DBU)

CAS Registry Number   6674-22-2

CAS Name   Pyrimido[1,2*a*]azepine, 2,3,4,6,7,8,9,10-octahydro-

Molecular Formula

Molecular Weight   152.24

Boiling Point   80-83°/0.6 mmHg

Melting Point   Not available.

Density 1.018 kg/m^3

Refractive Index   1.5219

Safety and Handling   Corrosive. Moisture sensitive.

Reactions   Base in elimination reactions. Reduction. Review: *Synthesis*, 1972, 591.

Availability   Aldrich: 96%, p.

Lancaster Synthesis: 97%, p.

Sigma: approx. 97%, p.

Horner-Emmons olefination of hydroxyoxoalkylphosphonates and related compounds

O. Tsuge*, S. Kanemasa, N. Nakagawa, H. Suga, *Bull. Chem. Soc. Jpn.*, 1987, **60**(*11*), 4091-4098

R = alkyl, PhCO$_2$Me
R'= Ph, alkyl

21 - 91 %
9 examples

Selenium-assisted reduction of aromatic ketones with carbon monoxide and water

Y. Nishiyama*, S. Hamanaka, A. Ogawa, N. Kambe, N. Sonoda*, *J. Org. Chem.*, 1988, **53**(*6*), 1326-1329

PhCMe $\xrightarrow[\text{120°, 24 h}]{\text{Se, CO, H}_2\text{O, DBU}}$ PhEt

95 %
13 examples

Facile synthesis of *S*-alkyl carbonothioates via direct *O*-carbonylation of alcohols

T. Mizuno*, I. Nishiguchi, T. Hirashima, A. Ogawa, N. Kambe, N. Sonoda, *Tetrahedron Lett.*, 1988, **29**(*37*), 4767-4768

n-BuOH $\xrightarrow[\text{ii) PhCH}_2\text{Br}]{\text{i) CO, sulphur, DBU, THF, 80°, 4 h}}$ n-BuOCOSCH$_2$Ph

86 %
12 examples

37. Diazomethane

CAS Registry Number	334-88-3
CAS Name	Methane, diazo-
Molecular Formula	CH_2N_2
Molecular Weight	42.04
Boiling Point	~0^oC
Melting Point	-145°C
Density	Not available.
Refractive Index	Not available.
Safety and Handling	Extreme risk of explosion by shock, friction, contact with ground glass surfaces, fire or other sources of ignition. Toxic by inhalation.
Reactions	Methylation. Methylenation. Reviews: *Aldrichim. Acta*, 1970, **3**(*4*), 9-12; 1983, **16**(*1*), 3-10.
Availability	Not commercially available.
Preparation	Aldrich supply a variety of kits with which to prepare diazomethane in quantities of ~100 mmol to larger scale (eg. 0.2-0.3 mol). The distillation glassware used with Diazald®(*N*-methyl-*N*-nitroso-*p*-toluenesulpho-namide) is supplied with nonground joints: the use of MNNG (1-methyl-3-nitro-1-nitrosoguanidine) as precursor allows the preparation of diazomethane without distillation. Further details: Aldrich catalogue, 1988-1989, pp. 2034-2036.

Synthesis of methyl acetamidoaryl butenoates in two highly selective steps

C. Cativiela*, M. D. Diaz de Villegas, E. Melendez, *Synthesis*, 1986, (5), 418-419

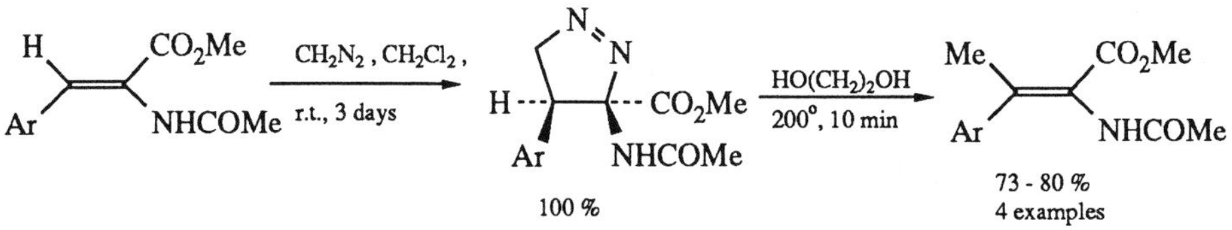
H
CO_2Me
Ar
NHCOMe
CH_2N_2, CH_2Cl_2,
r.t., 3 days
N
N
H
CO_2Me
Ar
NHCOMe
100 %
$HO(CH_2)_2OH$
200°, 10 min
Me
CO_2Me
Ar
NHCOMe
73 - 80 %
4 examples

Methylation of alcohols and phenols adsorbed on silica gel with diazomethane

H. Ogawa, T. Hagiwara, T. Chihara, S. Teratani*, K. Taya, *Bull. Chem. Soc. Jpn.*, 1987, **60**(2), 627-629

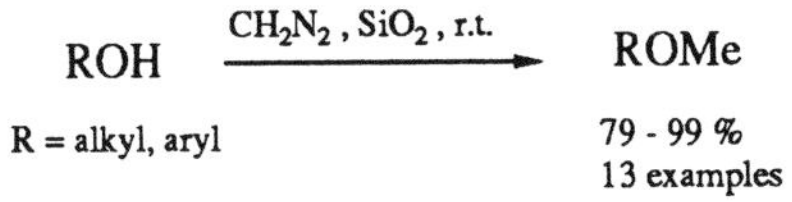
ROH
CH_2N_2, SiO_2, r.t.
ROMe
R = alkyl, aryl
79 - 99 %
13 examples

Oxiranes from methylenation of esters by diazomethane

P. Strazzolini, G. Verardo, A. G. Giumanini*, *J. Org. Chem.*, 1988, **53**(*14*), 3321-3325

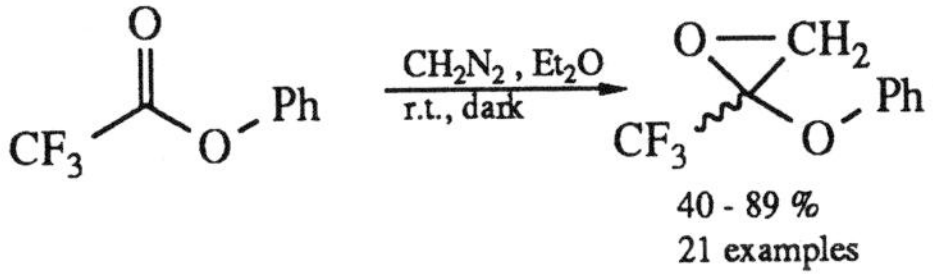
O
CF_3
O
Ph
CH_2N_2, Et_2O
r.t., dark
O
CH_2
Ph
CF_3
O
40 - 89 %
21 examples

38. 2,3-Dichloro-5,6-dicyanobenzoquinone (DDQ)

CAS Registry Number   84-58-2

CAS Name   1,4-Cyclohexadiene-1,2-dicarbonitrile, 4,5-dichloro-3,6-dioxo-

Molecular Formula

$$
\begin{array}{c}
\text{O} \\
\text{NC} \diagdown \quad \diagup \text{Cl} \\
\text{NC} \diagup \quad \diagdown \text{Cl} \\
\text{O}
\end{array}
$$

Molecular Weight   227.01

Boiling Point   Not available.

Melting Point   213-216°C

Density   Not available.

Refractive Index   Not available.

Safety and Handling   Harmful.

Reactions   Dehydrogenation reagent. Oxidation. Aromatization. Reviews: *Chem. Rev.*, 1967, **67**, 153; 1978, **78**, 317.

Availability   Aldrich: 98%, £.

Lancaster Synthesis: 98+%, £, bulk prices available.

Sigma: orange-yellow crystals, £.

Oxidative cyclization of (hydroxyalkyl)furans with DDQ

L. M. Harwood*, J. Robertson, *Tetrahedron Lett.*, 1987, **28**(*43*), 5175-5176

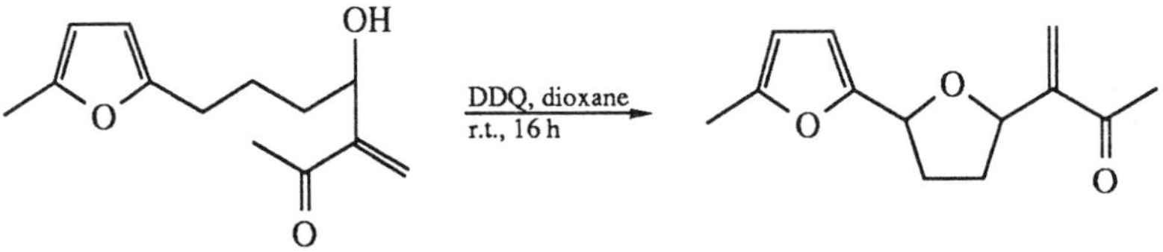

60 %
6 examples

Oxidative carbon-carbon bond formation by reaction of allyl ethers, silyl carbon nucleophiles and DDQ

Y. Hayashi, T. Mukaiyama, *Chem. Lett.*, 1987, (*9*), 1811-1814

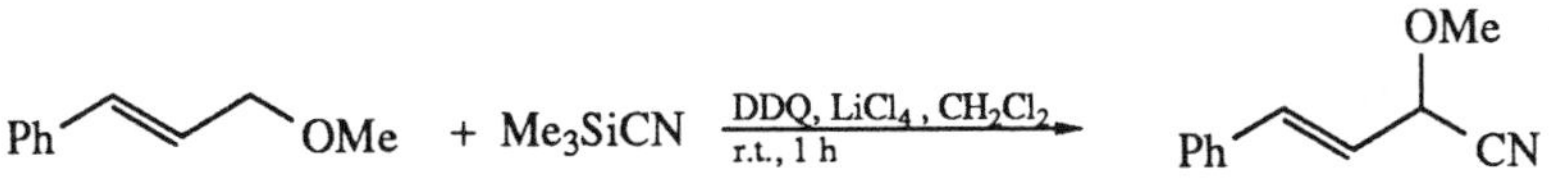

82 %
15 examples

DDQ in aqueous acetic acid, a convenient new reagent for the synthesis of aryl ketones and aldehydes via benzylic oxidation of arylalkanes

H. Lee, R. G. Harvey*, *J. Org. Chem.*, 1988, **53**(*19*), 4587-4589

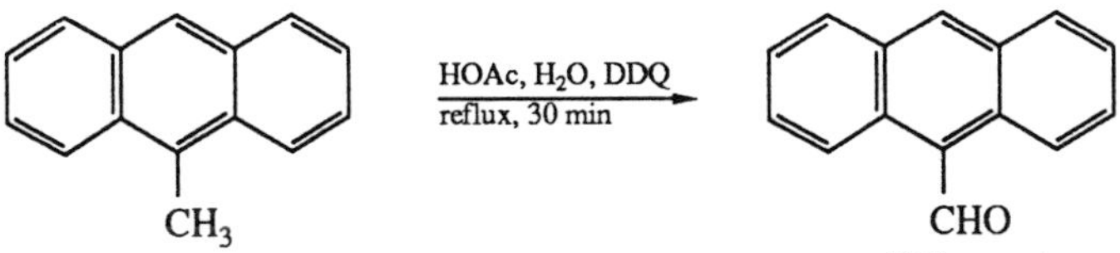

90 %
15 examples

39. Dicobalt octacarbonyl

CAS Registry Number	10210-68-1
CAS Name	Cobalt, di-μ-carbonylhexacarbonyldi-, (Co-Co)
Molecular Formula	[Co(CO)$_4$]$_2$
Molecular Weight	341.95
Boiling Point	52°C (dec.)
Melting Point	51°C
Density	1.73 kg/m^3
Refractive Index	Not available.
Safety and Handling	Air sensitive.
Reactions	Carbonylation.
Availability	Not commercially available.
Preparation	Prepared by treating CoCO$_3$ with carbon monoxide and hydrogen at 250 -300 atm and 120-200°C.
Other Preparations	*Magy. Asvanyolaj-Foldgaz Kiserl. Intez. Kozl.*, 1966, (7), 75-87.

Dicobalt octacarbonyl catalyzed conversion of benzylic alcohols to thiols, hydrocarbons and esters

H. Alper*, F. Sibtain, *J. Org. Chem.*, 1988, **53**(*14*), 3306-3309

$$C_6H_5CH_2OH \ + \ H_2S \ \xrightarrow[\text{CO, 60 atm}]{Co_2(CO)_8 \ , \ 150°} \ C_6H_5CH_2SH \ + \ C_6H_5CH_3$$

44 %　　　12 %

19 examples

Stereoselective siloxymethylation of glycosyl acetates

N. Chatani, T. Ikeda, T. Sano, N. Sonoda, H. Kurosawa, Y. Kawasaki, S. Murai*, *J. Org. Chem.*, 1988, **53**(*14*), 3387-3389

75 %
6 examples

Novel direct diacylation of Schiff bases

G. Vasapollo, H. Alper*, *Tetrahedron Lett.*, 1988, **29**(*40*), 5113-5116

$$PhCH=NPh \ + \ CO \ + \ MeI \ \xrightarrow[\text{CH}_2\text{Cl}_2 \, , \, 60°, \, 1 \, \text{atm}, \, 3 \, \text{h}]{\substack{Co_2(CO)_8 \, , \, KOH \\ \text{polyethylene glycol-400}}}$$

$$\underset{\substack{| \quad | \\ PhCHNPh}}{MeOC \quad COMe} \ + \ \underset{\substack{| \\ PhCH_2N-Ph}}{COMe}$$

76 %　　　5 %

18 examples

40. Diethylaluminium chloride

CAS Registry Number	96-10-6
CAS Name	Aluminum, chlorodiethyl
Molecular Formula	Et_2AlCl
Molecular Weight	120.56
Boiling Point	125-126°C/50 mmHg
Melting Point	-50°C
Density	0.961 kg/m^3
Refractive Index	Not available.
Safety and Handling	Pyrophoric. Moisture sensitive.
Reactions	Ring enlargement. Ene reaction catalyst. Cycloaddition. Reagent for aldol condensation: *J. Am. Chem. Soc.*, 1977, **99**, 7705.
Availability	Aldrich: under N_2 in Sure/Pac™ cylinders), p; 1M in hexanes under N_2 in Sure/Pac™ cylinders or Sure/Seal™ bottles, p; 1.8M in PhMe under N_2 in Sure/Seal™ cylinders or Sure/Seal™ bottles, p.

Ring expansion of ketones to α-methoxy and β-phenylthio ketones

B. M. Trost*, G. K. Mikhail, *J. Am. Chem. Soc.*, 1987, **109**(*13*), 4124-4127

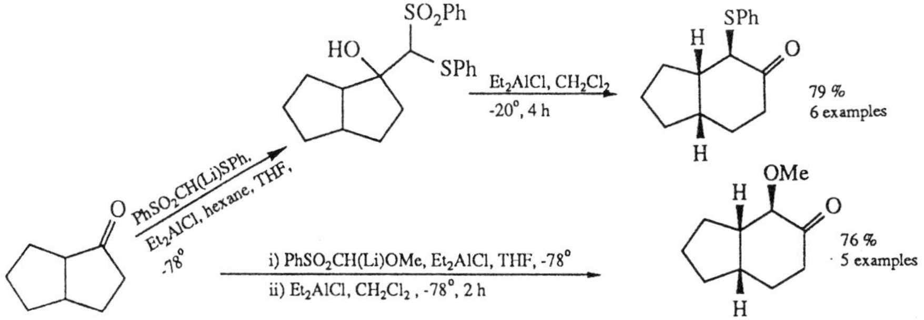

79 %
6 examples

76 %
5 examples

Dialkylaluminium chloride catalyzed ene reactions of carbonyl sulphide

L. V. Dunkerton*, M. Susa, *Synth. Commun.*, 1987, **17**(*10*), 1217-1225

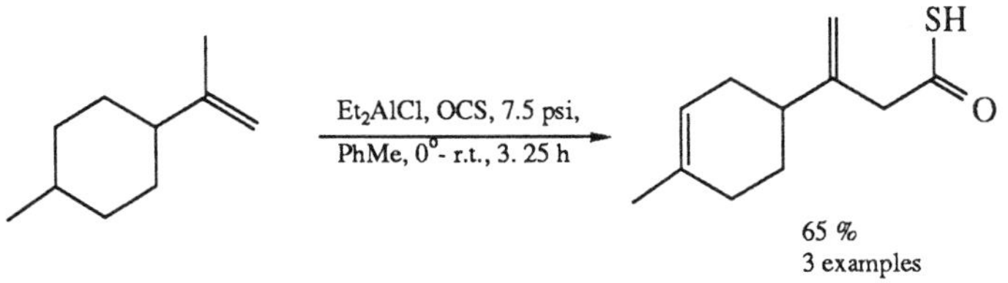

65 %
3 examples

Rapid and efficient cycloaddition of simple imines with activated dienes to give lactams

M. M. Midland*, J. I. McLoughlin, *Tetrahedron Lett.*, 1988, **29**(*37*), 4653-4656

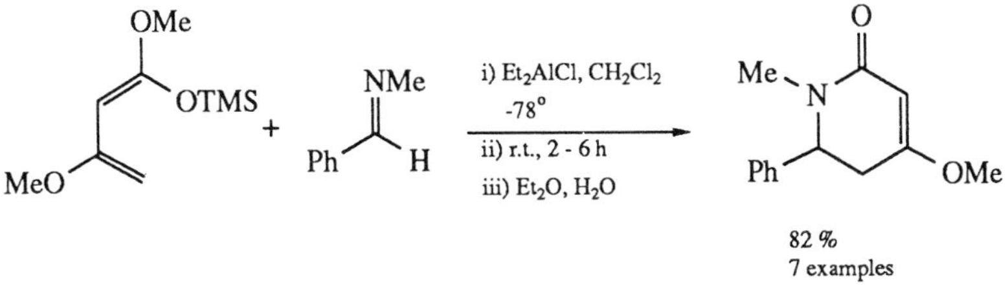

82 %
7 examples

41. Diethylaminosulphur trifluoride (DAST)

CAS Registry Number	38078-09-0
CAS Name	Sulfur, (N-ethylethanaminato)trifluoro-
Molecular Formula	Et_2NSF_3
Molecular Weight	161.19
Boiling Point	30-32°C/3 mmHg
Melting Point	Not available.
Density	1.220 kg/m^3
Refractive Index	Not available.
Safety and Handling	Corrosive. Moisture sensitive.
Reactions	Fluorinating agent prepared by reaction of Et_2NSiMe_3 with SF_4 in $FCCl_3$ or Et_2O at -78°: *Synthesis*, 1973, (*12*), 787-789; *J. Org. Chem.*; 1975, **40**(*5*), 574-578. Reviews: *Org. React.*, 1974, **21**, 1; 1985, **34**, 319.
Availability	Aldrich; ££.
	Lancaster Synthesis: ££, bulk prices available.
	Sigma: Inquire for details.

Preparation of glycosyl fluorides from monosaccharide hemiacetal using DAST

G. H. Posner*, S. R. Haines*, *Tetrahedron Lett.*, 1985, **26**(*1*), 5-8

87 % α, 13 % β
7 examples

Preparation of vicinal difluoroolefinic carbonyl compounds

A. E. Asato*, R. S. H. Liu*, *Tetrahedron Lett.*, 1986, **27**(*29*), 3337-3340

i) DAST, N-methyl-2-pyrrolidinone, -70°; r.t., 48 - 64 h
ii) NBS, (BzO)$_2$, CCl$_4$, reflux, 18 h

4 examples

Stereocontrolled access to and hydrogen fluoride abstraction from vicinal difluoroalkanes

T. Hamatani, S. Matsubara, H. Matsuda, M. Schlosser*, *Tetrahedron*, 1988, **44**(*10*), 2865-2874, 2875-2881

BuCH$_2$CHO

i) Me(CH$_2$)$_5$P$^+$Ph$_3$Br$^-$
NaNH$_2$, KOBu-t, Et$_2$O,
-75°, 1 h; 25°, overnight
ii) monoperphthalic acid, Et$_2$O
25°, 20 min

86 %

i) HF, Et$_3$N, 150°, 3 h
ii) DAST, CH$_2$Cl$_2$, -15°
pyridine; r.t., overnight
iii) KOBu-t, THF, 75°, 16 h

66 %

42. Diethyl azodicarboxylate (DEAD)

CAS Registry Number	1972-28-7
CAS Name	Diazenecarboxylic acid diethyl ester
Molecular Formula	$EtO_2CN=NCO_2Et$
Molecular Weight	174.16
Boiling Point	$106°C/13$ mm Hg
Melting Point	Not available.
Density	1.106 kg/m^3
Refractive Index	1.4220
Safety and Handling	Light sensitive.
Reactions	Dienophile. Oxidizing agent. Condensation agent.
Availability	Aldrich: 98%, £.
	Sigma: £.

Preparation of alkyl halides and cyanides from alcohols

S. Manna, J. R. Falck*, C. Mioskowski, *Synth. Commun.*, 1985, **15**(*8*), 663-668

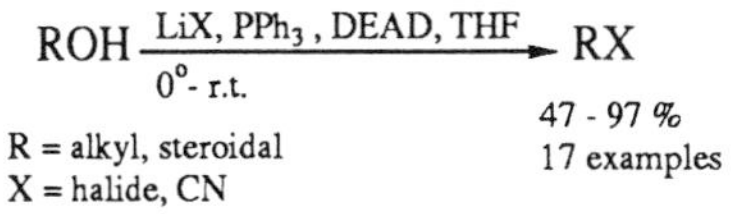
ROH
LiX, PPh3, DEAD, THF
0°- r.t.
RX
R = alkyl, steroidal
X = halide, CN
47 - 97 %
17 examples

A mild procedure for the preparation of 2-oxazolines

D. M. Roush*, M. M. Patel, *Synth. Commun.*, 1985, **15**(*8*), 675-679

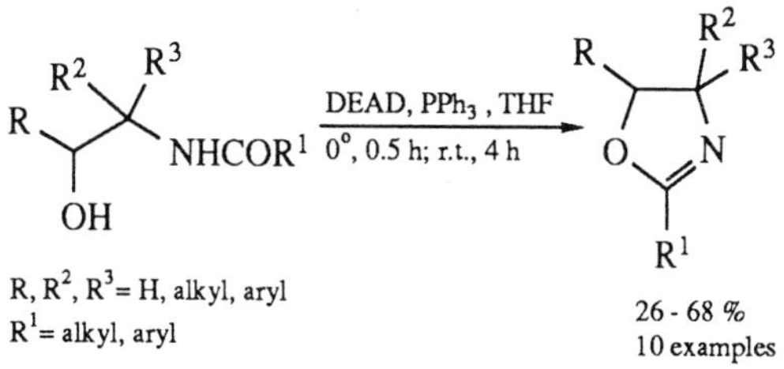
R
R2
R3
NHCOR1
OH
DEAD, PPh3, THF
0°, 0.5 h; r.t., 4 h
O
N
R1
R, R2, R3= H, alkyl, aryl
R1= alkyl, aryl
26 - 68 %
10 examples

A facile synthesis of diethyl (phenylthio)alkylphosphonates

T. Gajda, *Synthesis*, 1988, (*4*), 327-328

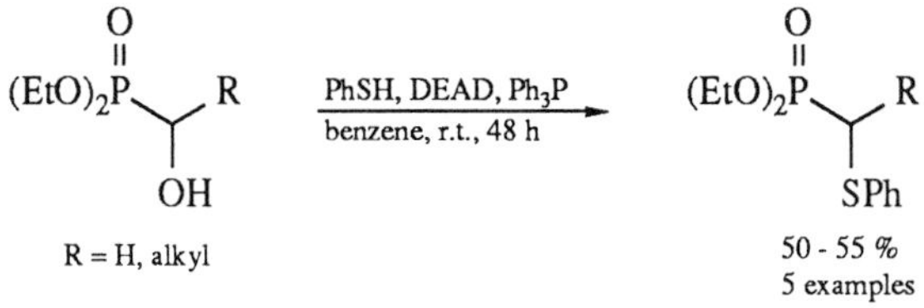
O
(EtO)2P
R
OH
PhSH, DEAD, Ph3P
benzene, r.t., 48 h
SPh
R = H, alkyl
50 - 55 %
5 examples

43. Diethylzinc

CAS Registry Number	557-20-0
CAS Name	Zinc, diethyl-
Molecular Formula	Et_2Zn
Molecular Weight	123.49
Boiling Point	117^oC
Melting Point	-28^oC
Density	1.205 kg/m^3
Refractive Index	1.4983
Safety and Handling	Pyrophoric. Corrosive. Reacts violently with water, CH_2Cl_2/alkene mixtures, MeOH, halogens or SO_2.
Reactions	Conjugate addition. Nucleophilic transfer of ethyl group. Useful reagent for preparation of carbenes in combination with diiodoalkanes.
Availability	Aldrich: p; 1M in hexanes, p; 1M in PhMe, p; all under N_2 in Sure/Pac™ cylinders or Sure/Seal™ bottles.

Asymmetric cyclopropanation via a Simmons-Smith reaction

I. Arai, A. Mori, H. Yamamoto*, *J. Am. Chem. Soc.*, 1985, **107**(*26*), 8254-8256

Ph $\diagdown$ O $\diagup$ CO$_2$Et

$\xrightarrow[\text{-20}^\circ\text{, 6 h; 0}^\circ\text{, 6 h}]{\text{Et}_2\text{Zn, CH}_2\text{I}_2\text{ , hexane}}$

91 % yield
87 % d.e.
10 examples

A convenient synthesis of vinyl sulphides

A. D. Rodriguez, A. Nikon*, *Tetrahedron*, 1985, 41(*20*), 4443-4448

SPh
SPh $\xrightarrow[\text{70}^\circ\text{, 1 h}]{\text{Et}_2\text{Zn, CH}_2\text{I}_2}$ SPh

92 %
73:27 mixture of isomers
6 examples

Syntheses of *trans*-β-lactams by reaction of α-imino esters with diethylzinc

M. R. P. van Vliet, J. T. B. H. Jastrzebski, W. J. Kaver, K. Goubitz, G. van Koten*, *Recl. Trav. Chim. Pays-Bas*, 1987, **106**(*4*), 132-134

R'O
$\xrightarrow[\text{-80}^\circ]{\text{Et}_2\text{Zn, pentane}}$
$\xrightarrow[\text{ii) H}_2\text{O}]{\text{i) -80}^\circ\text{ to r.t.}}$

R, R' = alkyl

80 - 90 %
4 examples

86

44. Diisobutylaluminium hydride (DIBAL)

CAS Registry Number	1191-15-7
CAS Name	Aluminium, hydrobis(2-methylpropyl)-
Molecular Formula	$(Me_2CHCH_2)_2AlH$
Molecular Weight	142.22
Boiling Point	116-118°C/1 mmHg
Melting Point	Not available.
Density	0.798 kg/m^3
Refractive Index	Not available.
Safety and Handling	Pyrophoric. Moisture sensitive.
Reactions	Reduction. Hydroalumination. Review: K. Maruoka, H. Yamamoto, *Angew. Chem. (Int. Ed. Engl.)*, 1985, **24**(*8*), 668-682.
Availability	Aldrich: p; 1M in PhMe, p; all under N_2 in Sure/Pac™ cylinders, Sure/Seal™ bottles or Kilo-Lab™ cylinders. Kilo-Lab cylinders require deposit. Also available in cyclohexane, CH_2Cl_2, heptane, hexanes, THF.

Enantioselective synthesis of dialkyl alcohols

D. F. Taber*, P. B. Deker, M. D. Gaul, *J. Am. Chem. Soc.*, 1987, **109**(*24*), 7488-7494

method A: Zn(BH$_4$)$_2$, ZnCl$_2$, PhMe, -78°, 0.5 - 3 h   : 94 % yield, 92:8 mixture

method B: DIBAL, PhMe, -78°, 0.5 h; -65 to -60°, 1.5 h : 82 % yield, 4:96 mixture

1-Np = 1-naphthyl

Cleavage of oxathiolanes with diisobutylaluminium hydride

T. K. Kiladze, B. A. Kirilyuk, I. A. Mel'nitskii, V. V. Duoryanchikov, E. A. K. Kantor, D. L. Rakhmankulov, *Chem. Heterocycl. Compd. (Engl. Transl.)*, 1987, **23**(*4*), 471

i) DIBAL, PhMe, reflux, 1 - 10 h
ii) HCl, H$_2$O

HOCH$_2$CH$_2$SCHRR'

85 - 90 %

R, R'= H, Me

Generation of [α-(alkoxycarbonyl)vinyl]aluminium and aluminium allenolates by the hydroalumination of α,β-acetylenic carbonyl compounds and their reaction with carbonyl compounds

T. Tsuda*, T. Yoshida, T. Saegusa*, *J. Org. Chem.*, 1988, **53**(*5*), 1037-1040

68 - 90 %
7 examples

23 - 31 %
2 examples

88

45. Diisopinocampheylborane

CAS Registry Number   21932-54-7

CAS Name   Borane, bis(2,6,6-trimethylbicyclo[3.1.1]hept-3yl)-

Molecular Formula

Molecular Weight   286.31

Boiling Point   Not available.

Melting Point   Not available.

Density   Not available.

Refractive Index   Not available.

Safety and Handling   Not available.

Reactions   Asymmetric hydroboration. Oxidation.
Reviews: D. S. Matteson, *Synthesis*, 1986, (*12*), 973-985; M. Srebnik, P. V. Ramachandran, *Aldrichimica Acta*, 1987, **20**(*1*), 9-24; H. C. Brown, B. Singaram, *Acc. Chem. Res.*, 1988, **21**(*8*), 287-293.

Availability   Not commercially available.

Preparation   H. C. Brown, G. Zweifel, *J. Am. Chem. Soc.*, 1961, **83**, 486.

Synthesis of chiral boronic and borinic esters via asymmetric hydroboration of alkenes

H. C. Brown*, P. K. Jadhav, M. C. Desai, *Tetrahedron*, 1984, **40**(8), 1325-1332

i) 2-butene, THF, -25°, 12 h
ii) 1-pentene, -25°, 1 h
iii) MeCHO, r.t., 1 h

75 %
73 % e.e.

67 %
99 % e.e.

5 examples

Asymmetric synthesis of cyclohexenyl alkanols via a stereochemically stable allyl borane

H. C. Brown*, P. K. Jadhav, K. S. Bhat, *J. Am. Chem. Soc.*, 1985, **107**(8), 2564-2565

Ipc_2BH, -25°, 12 h

Ipc = isopinocampheyl

92 % selectivity

i) MeCHO, -78°, 6 h; -78 to 25°
ii) $HO(CH_2)_2NH_2$, pentane
0 - 25°, 1 h

64 %
94 % e.e.

Asymmetric hydroboration of heterocyclic alkenes with diisopinocampheylborane

H. C. Brown*, J. V. N. V. Prasad, *J. Am. Chem. Soc.*, 1986, **108**(8), 2049-2054

i) (+)-Ipc_2BH, -25°, 6 h
ii) MeCHO
iii) NaOH, H_2O_2

(S)- 87 %
100 % e.e
13 examples

46. 4-Dimethylaminopyridine (DMAP)

CAS Registry Number   1122-58-3

CAS Name   4-Pyridinamine, *N,N*-dimethyl-

Molecular Formula

Me~N~Me

(structure of 4-dimethylaminopyridine)

Molecular Weight   122.17

Boiling Point   Not available.

Melting Point   108-110°C

Density   Not available.

Refractive Index   Not available.

Safety and Handling   Highly toxic. Corrosive. Irritant.

Reactions   Hypernucleophilic acylation catalyst for hindered alcohols.
Reviews: *Angew. Chem.*, 1978, **90**(*8*), 602; *Angew. Chem. (Int. Ed. Engl.)*; 1978, **17**, 569; *Chem. Soc. Rev.*, 1983, **12**, 129.

Availability   Aldrich: 99%, p.

Lancaster Synthesis: 99%, p, bulk prices available.

Sigma: off-white to yellow crystals, £.

A very practical new method for macrolactonization

E. P. Boden, G. E. Keck*, *J. Org. Chem.*, 1985, **50**(*13*), 2394-2395

$$HO-(CH_2)_{14}CO_2H \xrightarrow[\text{THF, reflux, 16 h}]{\text{DCC, DMAP.HCl, CHCl}_3}$$

95 %
4 examples

Acylation of tetramic acids

K. Hori, M. Arai, K. Nomura, E. Yoshii*, *Chem. Pharm. Bull.*, 1987, **35**(*10*), 4368-4371

$$\xrightarrow[\substack{\text{CH}_2\text{Cl}_2, 0°, 10 \text{ min;} \\ \text{r.t., 1.5 h}}]{R^3CO_2H, \text{ DCC, DMAP}}$$

$$\xrightarrow[\text{r.t., 4.5 h}]{\text{Et}_3\text{N}, 0°, 10 \text{ min;}}$$

R^1, R^2= H, alkyl
R^3= alkyl, vinyl

37 - 95 %
17 examples

Transesterification of oxo esters with allyl alcohols

J. C. Gilbert*, T. A. Kelly, *J. Org. Chem.*, 1988, **53**(*2*), 449-450

$$\xrightarrow[\text{12 - 36 h}]{\substack{\text{DMAP, PhMe} \\ \text{4A mol. sieves, reflux}}}$$

R^1- R^4= H, Me

32 - 96 %
9 examples

47. Dimethyl sulphoxide (DMSO)

CAS Registry Number	67-68-5
CAS Name	Methane, sulfinylbis-
Molecular Formula	Me_2SO
Molecular Weight	78.13
Boiling Point	189^oC
Melting Point	18.4^oC
Density	1.101 kg/m^3
Refractive Index	1.4787
Safety and Handling	Hygroscopic. Violent reaction with oxidants, active halogen compounds, metal hydrides. Harmful if swallowed. Irritating to eyes.
	Fp 95^oC
Reactions	Oxidizing agent. Solvent for many inorganic ions. Review: *Bull. Soc. Chim. Fr.*, 1965, 1021. Forms, on reaction with NaH, dimsyl sodium, strong base for conversion of phosphonium salts to phosphoranes.
Availability	Aldrich: 99.9%, HPLC grade (800 ml unit in Sure-Seal™ bottle), p; 99.9% spectrophotometric grade, p; anhydrous, 99+%, under N_2 in Sure/Seal™ bottles, p; 99+%, p. Also available as $(CD_3)_2SO$, 100.0 atom % D, £££; 99.9 atom % D (contains 0.03% v/v TMS: useful for FT-NMR work), £.
	Lancaster Synthesis: 99+%, p, bulk prices available.
	Sigma: 99+%, p; ACS reagent, p.

Oxidation of vicinal diols to α-dicarbonyl compounds by trifluoroacetic anhydride-activated DMSO

C. M. Amon, M. G. Banwell*, G. L. Gravatt, *J. Org. Chem.*, 1987, **52**(*22*), 4851-4855

i) DMSO-$(F_3CCO)_2O$, CH_2Cl_2
-60°, 1.5 h
ii) Et_3N, -60°, 1.5 h

68 %
11 examples

The use of phenyl dichlorophosphate as an activating agent in the Pfitzner-Moffat oxidation of alcohols

H.-J. Liu*, J. M. Nyangulu, *Tetrahedron Lett.*, 1988, **29**(*26*), 3167-3170

DMSO, PDCP, Et_3N
CH_2Cl_2, -10 to +20°

PDCP = phenyldichlorophosphate

84 %
11 examples

Iodine-DMSO: a useful reagent for the conversion of 2-hydroxydibenzoylmethanes into flavones

J. K. Makrandi, V. Kumari, *Chem. Ind. (London)*, 1988, (*19*), 630

I_2, DMSO
70 - 80°, 30 min

90 %
6 examples

48. Diphenyl diselenide

CAS Registry Number	1666-13-3
CAS Name	Diselenide, diphenyl
Molecular Formula	Ph_2Se_2
Molecular Weight	312.13
Boiling Point	Not available.
Melting Point	61-63°C
Density	1.557 kg/m^3
Refractive Index	Not available.
Safety and Handling	Highly toxic. Stench.
Reactions	Reagent for introduction of unsaturation. Reviews: *Tetrahedron*, 1978, **34**, 1049; *Acc. Chem. Res.*, 1979, **12**, 22; 1984, **17**, 28.
Availability	Aldrich: 99%, ££.
	Lancaster Synthesis: 98%, ££, bulk prices available.

Electrochemical phenylselenoetherification of enols

M. L. Mihailovic*, S. Konstantinovic, R. Vukicevia, *Tetrahedron Lett.*, 1987, **28**(*37*), 4343-4346

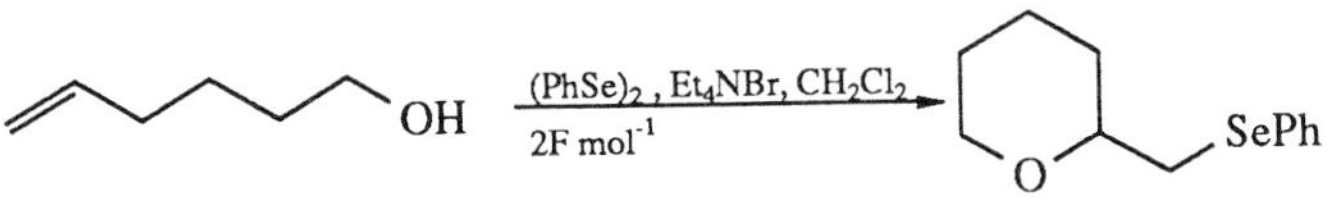

75 %
13 examples

Practical large-scale oxidation of hydroquinones to benzoquinones with hydrogen peroxide/diphenyl diselenide

D. V. Pratt, F. Ruan, P. B. Hopkins*, *J. Org. Chem.*, 1987, **52**,(*22*), 5053-5055

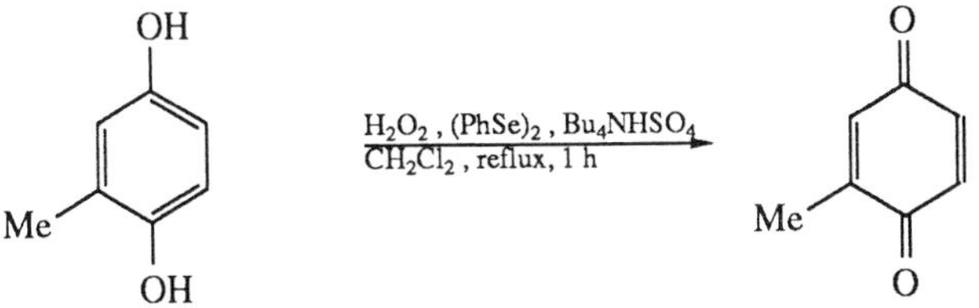

82 %
7 examples

Cleavage of cyclic ethers by phenylselenide anion

K. Haraguchi, H. Tanaka, H. Hayakawa, T. Miyasaka*, *Chem. Lett.*, 1988, (*6*), 931-934

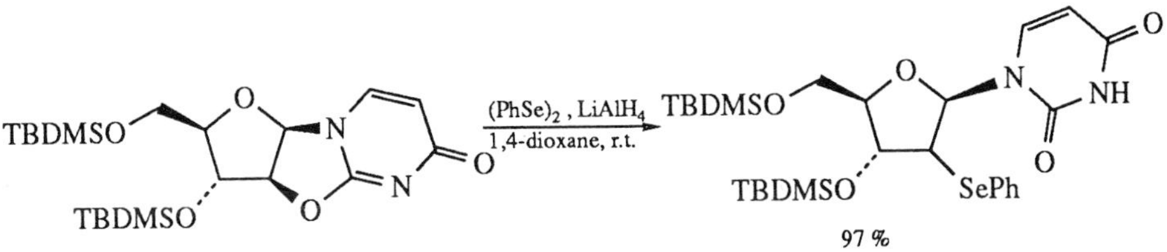

97 %
10 examples

96

49. Diphosphorus tetraiodide

CAS Registry Number	13455-00-0
CAS Name	Hypodiphosphorous tetraiodide
Molecular Formula	I_2PPI_2
Molecular Weight	569.57
Boiling Point	Not available.
Melting Point	$125^{\circ}C$
Density	Not available.
Refractive Index	Not available.
Safety and Handling	Corrosive. Light sensitive.
Reactions	Reagent for regioselective synthesis of alkyl iodides from alcohols. Reviews: H. Suzuki, H. Tani, *J. Synth. Org. Chem. Jpn.*, 1985, **43**(1), 76-83; *CA*, 203975t, Application of, in organic synthesis.
Availability	Aldrich: (under Ar), ££.

Reductive cleavage of aromatic azido, azo, azoxy and hydrazo compounds with diphosphorus tetraiodide

H. Suzuki*, H. Tani, S. Ishida, *Bull. Chem. Soc. Jpn.*, 1985, **58**(*6*), 1861-1862

ArN_3

$ArCON_3$

$ArSO_2N_3$

i) P_2I_4 , benzene, reflux,
2 - 50 h

ii) H_2O

$ArNH_2$

$ArCONH_2$ 13 - 86 %
15 examples

$ArSO_2NH_2$

Synthesis of furans from enediones by diphosphorus tetraiodide

S. H. Demirdji, M. J. Haddadin, C. H. Issidorides*, *J. Heterocyclic Chem.*, 1985, **22**(*2*), 495-496

P_2I_4 , $CHCl_3$, r.t., 20 - 30 min

55 - 76 %
6 examples

R = aryl, H

Hydrolysis of alkoxymethyl aryl ethers to give hydroxy arenes

H. Saimoto, Y. Kusano, T. Hiyama*, *Tetrahedron Lett.*, 1986, **27**(*14*), 1607-1610

ArOCHR'OR

P_2I_4 , CH_2Cl_2 , 0°, 25 min;
r.t., 5 min

R'= H, Me
R = alkyl

ArOH

56 - 92 %
7 examples

50. Hexabutylditin (Bis(tributyltin))

CAS Registry Number	813-19-4
CAS Name	Distannane, hexabutyl-
Molecular Formula	[Me(CH$_2$)$_3$]$_3$Sn)$_2$
Molecular Weight	580.08
Boiling Point	197-198°C/10 mmHg
Melting Point	Not available.
Density	1.148 kg/m^3
Refractive Index	1.5120
Safety and Handling	Moisture sensitive. Toxic.
Reactions	For the photodesulphurization of 1,3-dithiole-2-thiones to tetrathiafulvalenes: *J. Am. Chem. Soc.*, 1976, **98**, 7440. For preparation of tributylstannyllithiums. Reviews: *Chem. Rev.*, 1960, **60**, 459; *Chem. Ind. (London)*, 1972, 490.
Availability	Aldrich: 97%, £.
	Lancaster Synthesis: 98+%, £, bulk prices available.

Palladium-catalyzed coupling reaction of α-bromo ketones with hexabutylditin

M. Kosugi*, M. Koshiba, H. Sano, T. Migata*, *Bull. Chem. Soc. Jpn.*, 1985, 58(3), 1075-1076

R, R^1 = H, Me
R^2 = alkyl, Ph, OEt

49 - 81 %
8 examples

Reaction of hydroximic chlorides with hexabutylditin: generation and cycloaddition of nitrile oxides

B. H. Kim, *Synth. Commun.*, 1987, 17(10), 1199-1206

R = Ph, Et, t-Bu
R^1, R^2 = H, alkyl, aryl

39 - 92 %
14 examples

Atom transfer cycloaddition. Methylenecyclopentane synthesis

D. P. Curran*, M.-H. Chen, *J. Am. Chem. Soc.*, 1987, 109(21), 6558-6560

R = SiMe$_3$, H
R^1 = H, alkyl
R^2 = CO, CN, SO$_2$Ph

E:Z = 1. 2 - 10:1

exo:endo 7:1 - 1:0
total yield 27 - 68 %

51. Hydroxylamine-O-sulphonic acid

CAS Registry Number   2950-43-8

CAS Name   Hydroxylamine-O-sulfonic acid

Molecular Formula H_2NOSO_3H

Molecular Weight   113.09

Boiling Point   Not available.

Melting Point $210^{\circ}C$

Density   Not available.

Refractive Index   Not available.

Safety and Handling   Corrosive. Hygroscopic. Irritant. Keep cold.

Reactions   Versatile reagent for amination, hydroxymethylation. Reviews: *Aldrichim. Acta*, 1980, **13**, 3; *Org. Prep. Proced. Int.*, 1982, **14**, 265.

Availability   Aldrich: 97%, p; tech. grade, 90%, p.

Lancaster Synthesis: p, bulk prices available.

Sigma: light tan crystals, approx. 95%, p.

Preparation of primary sulphonamides by reaction of sulphinic acid salts with hydroxylamine-*O*-sulphonic acid

S. L. Granam, T. H. Scholz, *Synthesis*, 1986, (*12*), 1031-1032

$$PhSO_2Na \xrightarrow[\text{r.t., 16 h}]{\text{H}_2\text{NOSO}_3\text{H, NaOAc, H}_2\text{O}} PhSO_2NH_2$$

93 %
7 examples

Synthesis of amines from alkenes via organoboranes

H. C. Brown*, K.-W. Kim, M. Srebnik, B. Singaram, *Tetrahedron*, 1987, **43**(*18*), 4071-4078

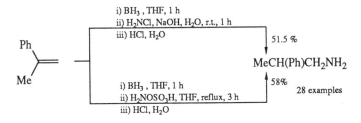

i) BH₃ , THF, 1 h
ii) H₂NCl, NaOH, H₂O, r.t., 1 h
iii) HCl, H₂O

51.5 %

MeCH(Ph)CH₂NH₂

i) BH₃ , THF, 1 h
ii) H₂NOSO₃H, THF, reflux, 3 h
iii) HCl, H₂O

58%
28 examples

Improved synthesis of ¹⁴C-labelled carboxylic acids from ¹⁴C-labelled amino acids by reaction with hydroxylamine-*O*-sulphonic acid

T. V. Ramamurthy*, S. Ravi, K. V. Viswanathan, *J. Labelled Compd. Radiopharm.*, 1988, **25**(*8*), 809-814

$$^{14}\text{C-PhCH}_2\text{CH(NH}_2)\text{CO}_2\text{H} \xrightarrow[\substack{\text{H}_2\text{O:MeOH (60:40)} \\ \Delta, \text{30 min}}]{\text{H}_2\text{NOSO}_3\text{H, KOH,}} {}^{14}\text{C-PhCH}_2\text{CH}_2\text{CO}_2\text{H}$$

95 %
11 examples

52. Iodosylbenzene

CAS Registry Number	536-80-1
CAS Name	Benzene, iodosyl-
Molecular Formula	$(PhIO)_n$
Molecular Weight	220.01 (PhIO)
Boiling Point	Not available.
Melting Point	Not available.
Density	Not available.
Refractive Index	Not available.
Safety and Handling	Explodes at $210^{\circ}C$.
Reactions	Oxidation. Reviews: A. Varvoglis, *Synthesis*, 1984, (*9*), 709-726; R. M. Moriarty, O. Prakash, *Acc. Chem. Res.*, 1986, **19**(*8*), 244-250.
Availability	Not commercially available.
Preparation	Prepared by reaction of iodosylbenzene diacetate with 3N sodium hydroxide under vigorous stirring followed by trituration of the solid formed: *Org. Synth., Coll. Vol. 5*, 658.

Hypervalent iodine oxidation of silyl enol ethers to α-hydroxy ketones

R. M. Moriarty*, M. P. Duncan, O. Prakash, *J. Chem. Soc., Perkin Trans. 1*, 1987, (8), 1781-1784

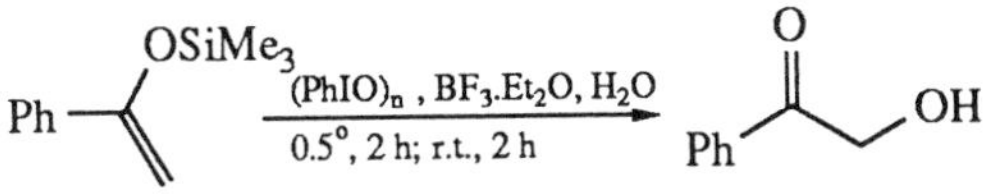

Conversion of thiocarbonyl into carbonyl in uracil, uridine, and *Escherichia coli* **tRNA by hypervalent iodine oxidation**

R. M. Moriarty*, I. Prakash, D. E. Clarisse, R. Penmasta, A. K. Awasthi, *J. Chem. Soc., Chem. Commun.*, 1987, (*16*), 1209-1210

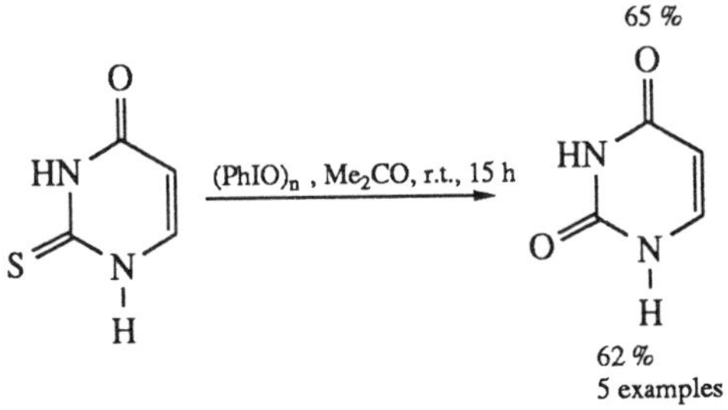

Allylation of aromatic compounds with allyltrimethylsilane using a hypervalent organoiodine compound

K. Lee, D. Y. Kim, D. Y. Oh*, *Tetrahedron Lett.*, 1988, 29(*6*), 667-668

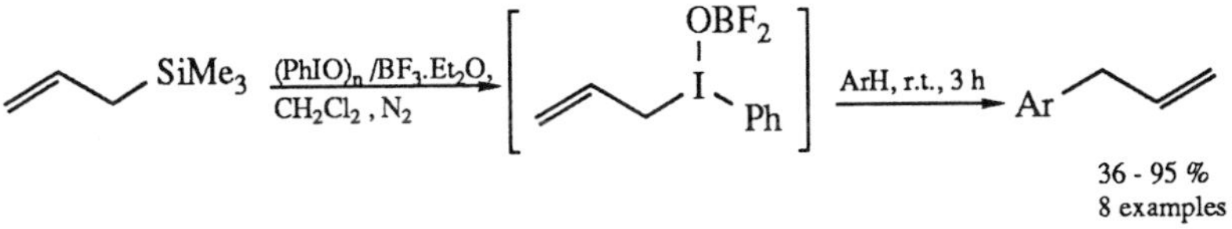

53. Iodotrimethylsilane

CAS Registry Number	16029-98-4
CAS Name	Silane, iodotrimethyl-
Molecular Formula	Me$_3$SiI
Molecular Weight	200.10
Boiling Point	106°C
Melting Point	Not available.
Density	1.406 kg/m^3
Refractive Index	1.4710
Safety and Handling	Flammable liquid. Corrosive. Keep cold. Fp -31°C
Reactions	Efficient reagent for ether and ester cleavage under neutral conditions. Reviews: *Aldrichim. Acta*, 1981, **14**(2), 31; *Synthesis*, 1980, 861; *Tetrahedron*, 1982, **38**, 2225.
Availability	Aldrich: 97%, £.
	Lancaster Synthesis: 98+%, £, bulk prices available.
	Sigma: 90-95%, £. Stabilized with copper.

Reduction of benzylic alcohols

T. Sakai*, K. Miyata, M. Utaka, A. Takeda, *Tetrahedron. Lett.*, 1987, **28**(*33*), 3817-3818

$$\underset{\substack{R^1 = H, Me \\ R^2 = H, alkyl}}{\overset{OH}{\underset{R^2}{Ar \diagdown R^1}}} \xrightarrow[\text{r.t., 24 h}]{Me_3SiCl, NaI (=Me_3SiI), MeCN, hexane} \underset{\substack{57 - 99 \% \\ 12 \text{ examples}}}{\overset{H}{\underset{R^2}{Ar \diagdown R^1}}}$$

Trimethylsilyl iodide-catalyzed reductive coupling of carbonyl compounds with trialkylsilanes to give ethers

M. B. Sassaman, K. D. Kotian, G. K. S. Prakash, G. A. Olah*, *J. Org. Chem.*, 1987, **52**(*19*), 4314-4319

R, R^1 = H, alkyl, aryl
R^2 = aryl

Me$_3$SiOTf /Et$_3$SiH, CH$_2$Cl$_2$, r.t., 2 h
or Me$_3$SiI /Et$_3$SiH, CH$_2$Cl$_2$, 0°, 10 min

quant.

R^{2}OSiMe$_3$, Me$_3$SiI, CH$_2$Cl$_2$,0°, 10 min

quant.
14 examples

Stereospecific functionalization of thiazolidine via silicon Pummerer reaction of thiazolidine S-oxides

N. Tokitoh, Y. Igurashi, W. Ando*, *Tetrahedron Lett.*, 1987, **28**(*47*), 5903-5906

R^1, R^2 = H, alkyl, aryl

t-BuMe$_2$SiOTf, CH$_2$Cl$_2$
NEt$_3$, 0°

27 - 72 %   0 - 50 %   8 examples

Me$_3$SiI, CH$_2$Cl$_2$,
NEt$_3$ or EtNPr-i, 0°, several h

42 - 54 %   0 - 20 %

R^1 = CH$_2$R^3

54. Iron pentacarbonyl

CAS Registry Number	13463-40-6
CAS Name	Iron, pentacarbonyl-
Molecular Formula	$Fe(CO)_5$
Molecular Weight	195.90
Boiling Point	103^oC
Melting Point	-20^oC
Density	$1.490 \ kg/m^3$
Refractive Index	1.5196
Safety and Handling	Highly toxic. Flammable liquid. Fp -15^oC
Reactions	Catalyst for addition, cycloaddition, reduction.
Availability	Aldrich: p.

β-Methylenecyclopentenones via iron carbonyl-induced [2+2+1]cycloaddition of alkynes to allenes and carbon monoxide

R. Aumann*, H.-J. Weidenhaupt, *Chem. Ber.*, 1987, **120**(*1*), 23-27

R, R^1= H, Ph
R^2= H, Ph, CH$_2$Ph

18 - 56 %
7 examples

Addition of trichloroalkanes to vinyltrimethylsilane in the presence of iron pentacarbonyl

A. A. Kamyshova, V. I. Dostrovalova, E. T. Chukhovskaya, *Bull. Acad. Sci. USSR, (Engl. Transl.)*, 1987, **36**(*5,2*), 1087-1089

$$RCCl_3 + H_2C=CHSiMe_3 \xrightarrow[\text{or i-PrOH, } 105 - 130°, 3 - 5 \text{ h}]{\text{Fe(CO)}_5, \text{PPh}_3, \text{DMF, HMPA}} RCCl_2CH_2CHClSiMe_3$$

R = alkyl

64 - 92 %
2 examples

Catalytic reduction of aryl iodides catalyzed by iron pentacarbonyl

J.-J. Brunet*, M. Taillifer, *J. Organomet. Chem.*, 1988, **348** (*1*),C5-C8

$$ArI \xrightarrow[\text{MeOH, } 60°, 2 - 48 \text{ h}]{\text{Fe(CO)}_5, \text{K}_2\text{CO}_3, \text{CO (1 atm)}} ArH$$

25 - 100 %

55. Lawesson's Reagent (*p*-Methoxyphenylthionophosphine sulphide dimer)

CAS Registry Number	19172-47-5
CAS Name	2,4-Bis(4-methoxyphenyl)-1,3-dithia-2,4-diphosphetane-2,4-disulphide
Molecular Formula	

$$\text{MeO} - \underset{}{\bigcirc} - \overset{S}{\underset{S}{P}} \overset{S}{\underset{S}{P}} - \bigcirc - \text{OMe}$$

Molecular Weight	404.47
Boiling Point	Not available.
Melting Point	227/228-229°C
Density	Not available.
Refractive Index	Not available.
Safety and Handling	Moisture sensitive. Stench.
Reactions	Thiation agent: *Bull. Soc. Chim. Belg.*, 1978, **87**(*4*), 223; 229; 293-297; 299; 525-534. Reviews: R. A. Cherkasov, G. A. Kutyrev, A. N. Pudovik, Tetrahedron, 1985,**41**(*13*), 2567-2624; **41**(*22*), 5061-5087.
Availability	Aldrich: 97%, p.
	Lancaster Synthesis: p, bulk prices available.
	Sigma: £.

Sulphuration of trialkyl phosphites and triphenylphosphine with Lawesson's reagent

N. G. Zabirov, R. A. Cherkasov, I. S. Khalikov, A. N. Pudovik, *J. Gen. Chem. USSR, (Engl. Transl.)*, 1986, **56**(*12,1*), 2365-2368

$$R_3P \xrightarrow[\text{reflux, 0.5 - 30 days}]{\text{Lawesson's reagent, benzene}} R_3P{=}S$$

R = alkoxy, Ph

45 - 90 %
4 examples

Reaction of acid chlorides with bis(methoxyphenyl)-dithiadiphosphetane disulphide: thioamidation of amines

N. M. Yousif, M. A. Salama, *Phosphorus Sulfur Relat. Elem.*, 1987, **32**(*1-2*), 51-53

RCOCl $\xrightarrow[110°, 1 h]{\text{Lawesson's reagent, PhMe,}}$

$\xrightarrow[\text{[R=Ph]}]{\text{NaOEt, PhMe, 40°, 1 h}}$

$$MeO{-}\!\!\bigcirc\!\!{-}\overset{\overset{S}{\|}}{\underset{\underset{OEt}{|}}{P}}{-}S{-}\overset{\overset{O}{\|}}{C}{-}Ph \quad 50\%$$

$\xrightarrow[70°, 2 - 4 \text{ min}]{\text{HNR}^1\text{R}^2, \text{Et}_3\text{N, PhMe,}}$

$$R{-}\overset{\overset{S}{\|}}{C}{-}N\overset{R^1}{\underset{R^2}{\diagdown}}$$

47 - 95 %
4 examples

R = Ph, Me
R^1, R^2 = H, alkyl, aryl

Thionation of pyrimidinediones with Lawesson's reagent

K. Kaneko*, H. Katayama, T. Wakabayashi, T. Kumonaka, *Synthesis*, 1988, (2), 152-154

$$\xrightarrow[120°, 1 h]{\text{Lawesson's reagent, HMPT}}$$

78 %
4 examples

56. Lead(IV) acetate (Lead tetraacetate)

CAS Registry Number   546-67-8

CAS Name   Acetic acid, compounds, lead(4+) salt

Molecular Formula $(AcO)_4Pb$

Molecular Weight   443.37

Boiling Point   Not available.

Melting Point $175^{\circ}C$

Density $2.228 \ kg/m^3$

Refractive Index   Not available.

Safety and Handling   Highly toxic. Moisture sensitive. Irritant.

Reactions   Oxidizing agent. Acetoxylation reagent.
Reviews: *Synthesis*, 1970, 279; 1971, 501; 1973, 567; *Org. React.*, 1972, **19**, 279.

Availability   Aldrich: p.

Lancaster Synthesis: p, bulk prices available.

Johnson Matthey: p.

Sigma: approx. 95%, p.

Synthesis of Z-allyl acetates from cyclic homoallyl alcohols

P. Ramaiah, A. S. Rao*, *Tetrahedron Lett.*, 1988, **29**(*17*), 2119-2120

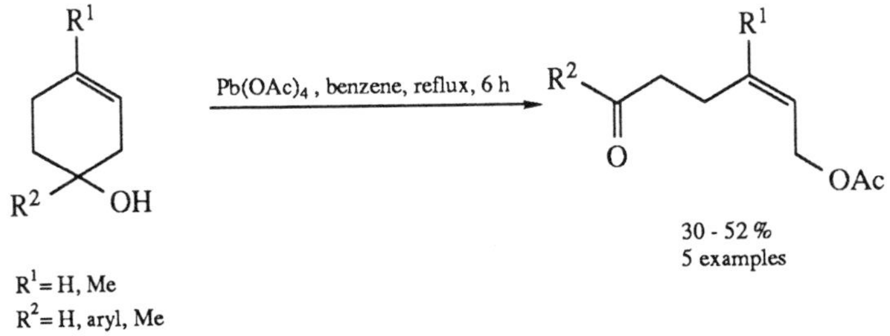
R1
R2
OH
Pb(OAc)4 , benzene, reflux, 6 h
R2
O
R1
OAc
30 - 52 %
5 examples
R1= H, Me
R2= H, aryl, Me

Oxidation of 1-hydroxyazetidines to β-lactams

P. A. Van Elburg, D. N. Reinhoudt, *Recl. Trav. Chim. Pays-Bas*, 1988, **107**(*5*), 381-387

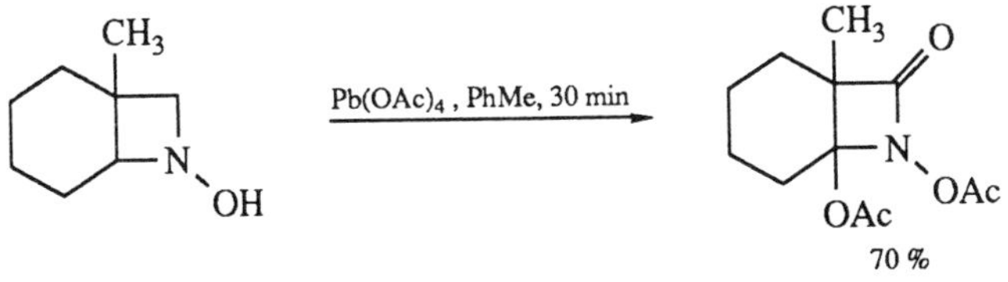
CH3
N
OH
Pb(OAc)4 , PhMe, 30 min
CH3
O
N
OAc
OAc
70 %

α-Vinylation of β-dicarbonyls using lead tetraacetate and divinyl mercury compounds

M. G. Moloney, J. T. Pinhey*, *J. Chem. Soc., Perkin Trans. 1*, 1988, (*10*), 2847-2854

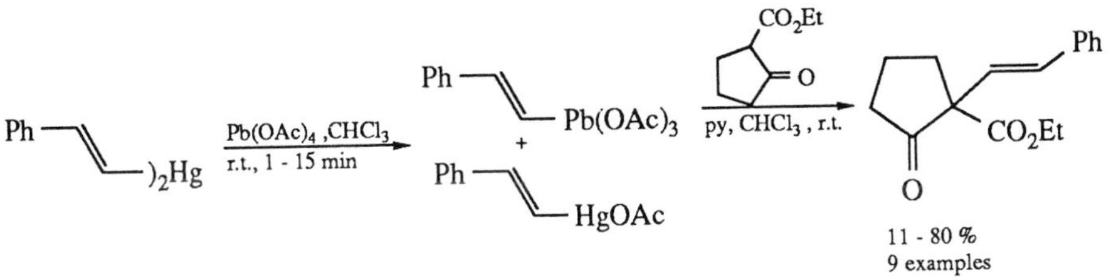
Ph
)2Hg
Pb(OAc)4 ,CHCl3
r.t., 1 - 15 min
Ph
Pb(OAc)3
+
Ph
HgOAc
CO2Et
O
py, CHCl3 , r.t.
Ph
CO2Et
O
11 - 80 %
9 examples

57. Lithium dimethylcuprate

CAS Registry Number	15681-48-8
CAS Name	Cuprate(1-), dimethyl-, lithium
Molecular Formula	Me$_2$CuLi
Molecular Weight	100.56
Boiling Point	Not available.
Melting Point	Not available.
Density	Not available.
Refractive Index	Not available.
Safety and Handling	Moisture sensitive. Air sensitive.
Reactions	Reductive cleavage. Cross-coupling with alkyl halides. Addition to double bonds. Alkene synthesis from allyl acetates: R. J. Anderson, C. A. Henrick, J. B. Siddal, *J. Am. Chem. Soc.*, 1970, **92**(*3*), 735-737. Reviews: B. H. Lipschutz, R. S. Wilhelm, J. A. Kozlowski, *Tetrahedron*, 1984, **40**(*24*), 5005-5038; B. H. Lipschutz, *Synthesis*, 1987, (*4*), 325-341 (higher-order cuprates); Y. Yamamoto, *Angew. Chem. (Int. Ed. Engl.)*, 1986, **25**(*11*), 947-959.
Availability	Not commercially available.
Preparation	R$_2$CuLi are prepared *in situ* by mixing 2 mol RLi with 1 mol cuprous halide in ether at low temperatures, or by dissolving an alkyl copper compound in an alkyllithium solution.
Other Preparations	An improved method is given by House *et al.*, *J. Org. Chem.*, 1975, **40**, 1460. See also: *An Introduction to Synthesis using Organocopper Reagents*, G. H. Posner, Interscience, NY, 1980.

Carbocupration of acetylenic acetals and ketals

A. Alexakis*, A. Commercon, C. Coulentianos, S. F. Normant, *Tetrahedron*, 1984, **40**(*4*), 715-731

$$R_2CuLi \ + \ 2 \ HC\equiv C-\overset{\overset{OR^2}{|}}{\underset{\underset{R^1}{|}}{C}}-OR^2 \ \xrightarrow[H_2O]{Et_2O}$$

R, H, OR², R¹, OR² structure

54 - 91 %

Alkylation of carbohydrate primary tosylates with organocuprate or Grignard reagents

J.-R. Pougny, *Tetrahedron Lett.*, 1984, **25**(*22*), 2363-2366

TsO, Me₃SiO, O, OMe structure
$\xrightarrow[-78°, \ 1 \ h; \ -5°, \ 2 \ h]{Me_2CuLi, \ C_6H_6 , \ Et_2O}$
Me, Me₃SiO, O, OMe structure

93 %
11 examples

Reductive cleavage of γ,δ-alkylidenedioxy-α,β-unsaturated esters promoted by organocuprates

S. Takano*, Y. Sekiguchi, K. Ogasawara, *J. Chem. Soc., Chem. Commun.*, 1988, (*7*), 449-450

R, H, H, CO₂R³ structure
$\xrightarrow[-70°]{Me_2CuLi, \ Et_2O}$
R, CO₂R³, R², R¹, OH, OH structure

27 - 94 %
6 examples

R - R² = H, alkyl,
R³ = alkyl
CO₂R³ = CN

58. Lithium naphthalenide

CAS Registry Number   7308-67-0

CAS Name   Naphthalene, radical ion (1-), lithium

Molecular Formula

$$\left[\bigcirc\!\!\bigcirc\right]^{-\cdot} Li^+$$

Molecular Weight   134.11

Boiling Point   Not available.

Melting Point   Not available.

Density   Not available.

Refractive Index   Not available.

Safety and Handling   Moisture sensitive. Air sensitive.

Reactions   Catalyses reaction of amines with alkenes: *J. Chem. Technol. Biotechnol.*, 1987, **37**(2), 95-99.

Availability   Not commercially available.

Preparation   T. Azuma, S. Yanagida, H. Sakurai, S. Sasa, K. Yoshino, *Synth. Commun.*, 1982, **12**(2), 137-140.

Other Preparations   *Chem. Ind. (London)*, 1983(*4*), 167-168; *Nippon Kagaku Kaishi*, 1984(*11*), 1744-1746.

One-pot synthesis of conjugated enynes

J. Barluenga*, M. Yus, J. M. Concellon, P. Bernad, F. Alvarez, *J. Chem. Res. Synop.*, 1985, (*4*), 128-129

$$R^1HC-\underset{\underset{Cl}{|}}{\overset{\overset{O}{||}}{C}}-R^2 \quad \begin{array}{l} \text{i) } R^3C\equiv CMgBr, Et_2O\text{-THF, Ar} \\ \quad 0°, 8\text{ h} \\ \text{ii) } LiC_{10}H_8, THF, -78°\text{- r.t.} \\ \hline \quad \text{overnight} \\ \text{iii) } H_3O^+ \end{array} \quad R^1CH=\underset{\underset{R^2}{|}}{C}-C\equiv CR^3$$

R, R^3= alkyl
R^2= H, alkyl

52 - 66 %
7 examples

Butyrolactones from carbonyl compounds and chloropropanal diethyl acetal

J. Barluenga*, J. R. Fernandez, M. Yus, *J. Chem. Soc., Chem. Commun.*, 1987, (*20*), 1534-1535

$$\text{CH}_2=\text{CH-CHO} \quad \xrightarrow{\text{HCl, EtOH}} \quad \underset{\underset{OEt}{|}}{Cl-CH_2CH_2-CH} \overset{OEt}{} \quad \begin{array}{l} \text{i) } LiC_{10}H_8, THF, -78°, 3\text{ h} \\ \text{ii) } R^1R^2CO, -78°\text{- r.t., overnight} \\ \text{iii) } NH_4Cl \\ \text{iv) } mCPBA, BF_3.Et_2O, CH_2Cl_2, \\ \quad \text{overnight} \end{array}$$

R^1, R^2= H, alkyl, aryl

44 - 66 %
7 examples

Rapid oxidative addition of copper to alkyl, aryl, alkynyl and vinyl halides

G. W. Ebert, R. D. Rieke*, *J. Org. Chem.*, *1988*, 53(*19*), 4482-4488

$$PhI \xrightarrow[10\text{ min, }25°]{CuI.PEt_3, LiC_{10}H_8, THF} PhCu \begin{array}{l} \xrightarrow{85°, 24\text{ h}} Ph\text{-}Ph \quad 66\% \\ \xrightarrow{H_2O} PhH \quad 98\% \\ \qquad\qquad 9\text{ examples} \end{array}$$

$$PhCH=CHBr \xrightarrow[10\text{ min, }0°]{CuI.PEt_3, LiC_{10}H_8, THF} PhCH=CHCu \xrightarrow{H_2O} PhCH=CH_2$$

82 %
6 examples

59. Mercury(II) acetate (Mercuric acetate)

CAS Registry Number	1600-27-7
CAS Name	Acetic acid, compounds, mercury(2+) salt.
Molecular Formula	$(AcO)_2Hg$
Molecular Weight	318.68
Boiling Point	Not available.
Melting Point	179-182°C
Density	3.270 kg/m^3
Refractive Index	Not available.
Safety and Handling	Highly toxic. Light sensitive.
Reactions	Mercuration reagent. Acetoxylation. Reviews: *Angew. Chem.*, 1978, **90**, 28; *Angew. Chem. (Int. Ed. Engl.)*, 1978, **17**, 27; *Tetrahedron*, 1982, **38**, 1713; *Synthesis*, 1973, 567.
Availability	Aldrich: 98+%, ACS reagent, p.
	Lancaster Synthesis: 98+%, p, bulk prices available.
	Johnson Matthey: crystalline, p.

Phase-transfer catalytic oxidation of terminal alkynes to keto aldehydes

F. P. Ballistreri, S. Failla, G. A. Tomaselli*, *J. Org. Chem.*, 1988, **53**(*4*), 830-831

$$PhC\equiv CH \xrightarrow[\substack{20^\circ,\ 30\ min}]{\substack{\text{Aliquat, Hg(OAc)}_2,\ \text{Na}_2\text{MoO}_4 \cdot 2\text{H}_2\text{O} \\ \text{H}_2\text{O}_2,\ \text{pH }3.5,\ \text{ClCH}_2\text{CH}_2\text{Cl},\ \text{H}_2\text{O}}} PhC\overset{\text{O}}{\overset{\|}{C}}CHO$$

43 %
2 examples

Alkylhalouracils via direct mercuration of alkyluracils

L. Skulski*, A. Kujiwa, T. M. Kujawa, *Bull. Pol. Acad. Sci., Chem.*, 1987, **35**(*11-12*), 499-505

i) Hg(OAc)$_2$, AcOH, Δ
ii) KI (aq.), Δ
iii) KI$_3$ (aq.), Δ

88 %
4 examples

Mercury(II) catalyzed one-pot regioselective synthesis of aminoethylfurans

J. Barluenga*, F. Aznar, M. Boyod, *Tetrahedron Lett.*, 1988, **29**(*39*), 5029-5032

R^1, R^2 = H, Me, Et, Ph
R^3, R^4 = morpholino, cycloalkyl

58 - 74 %
6 examples

118

60. Montmorillonite clay

CAS Registry Number   1318-93-0

CAS Name   Montmorillonite (($Al_{1.33-1.67}Mg_{0.33-0.67}$) $(Ca_{0-1}Na_{0-1})_{0.33}Si_4(OH)_2O_{10} \cdot xH_2O$)

Molecular Formula $(Al_{1.33-1.67}Mg_{0.33-0.67})(Ca_{0-1}Na_{0-1})_{0.33}$- $Si_4(OH)_2O_{10} \cdot xH_2O$

Molecular Weight   Not available.

Boiling Point   Not available.

Melting Point   Not available.

Density   Not available.

Refractive Index   Not available.

Safety and Handling   Not available.

Reactions   Michael reactions. Cycloadditions. Oxidations.

Availability   Aldrich: Montmorillonite K10, p; Montmorillonite KSF, p.

Allylic phenylation of hydroxy methylene esters with benzene and montmorillonite K10

D. Saib, A. Foucaud*, *J. Chem. Res., Synop.*, 1987, (*11*), 372-373

30 - 80 %
Z:E = 1:1.2 - 0:1
6 examples

R = alkyl, aryl

Synthesis of indoles from ketones and phenylhydrazines using montmorillonite clay as catalyst

P. Bhattacharyya*, S. S. Jash, *Indian J. Chem., Sect. B*, 1987, **26**(*12*), 1177

$$Me_2CO + PhNHNH_2 \xrightarrow[120°, 5\ min]{montmorillonite}$$

65 %
8 examples

Montmorillonite clay as an efficient heterogeneous catalyst for Michael reactions of silyl ketene acetals and silyl enol ethers with α,β-unsaturated carbonyl compounds

M. Kawai, M. Onaka*, Y. Izumi*, *Bull. Chem. Soc. Jpn.*, 1988, **61**(*6*), 2157-2164

Al-mont = aluminium ion exchanged montmorillonite

84 %
syn:anti = 27:73
19 examples

120

61. MoOPH
(Oxodiperoxymolybdenum-pyridine-hexamethylphosphoramide)

CAS Registry Number 23319-63-3

CAS Name Molybdenum, (hexamethylphosphoric triamide-O)-oxodiperoxy(pyridine)-

Molecular Formula

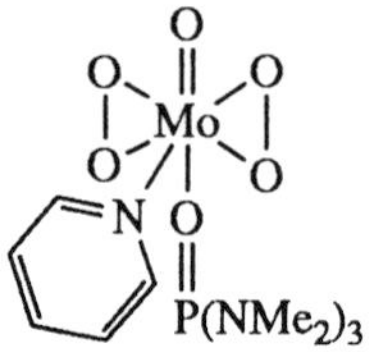

Molecular Weight 434.24

Boiling Point Not available.

Melting Point 116°C (dec.)

Density Not available.

Refractive Index Not available.

Safety and Handling Highly toxic. Cancer suspect agent.

Reactions Oxidation. Hydroxylation of enolates. Stereoselective oxidant for organoboranes: *J. Org. Chem.*, 1980, **45**, 4514.

Availability Aldrich: £££.

Asymmetric synthesis of α-hydroxy esters

R. Gamboni, P. Mohr, N. Waespe-Sarcevic, C. Tamm*, *Tetrahedron Lett.*, 1985, **26**(2), 203-206

O₂CCH₂CH₂Ph / OCH₂Ph → i) LICA, THF, -78° ii) MoOPH, -52 or -78°

O₂C—CH(OH)—CH₂Ph / OCH₂Ph + O₂C—CH(OH)—CH₂Ph / OCH₂Ph

LICA = lithium isopropylcyclohexylamide

80 %
2:8 mixture
4 examples

Trifluoromethyl group induced stereoselective synthesis of α-hydroxy carbonyl compounds

Y. Morizawa*, A. Yasuda, K. Uchida, *Tetrahedron Lett.*, 1986, **27**(*16*), 1833-1836

Me, F₃C—CO₂Et → i) LDA ii) MoOPH, THF, -78°; -20°, 3 h

Me—CH(OH)—CO₂Et / F₃C + Me—CH(OH)—CO₂Et / F₃C

75 %
93:3 mixture
6 examples

Synthesis of α-dicarbonyl compounds by oxidation of alkynes

F. P. Ballistreri, S. Failla, G. A. Tomeselli*, R. Curci, *Tetrahedron Lett.*, 1986, **27**(*42*), 5139-5142

$R-C \equiv C-R'$ → MoOPH, Hg(OAc)₂ / ClCH₂CH₂Cl, 40°, 0.25 - 20 h

$R-CO-CO-R'$

R, R'= H, alkyl, aryl

55 - 90%
8 examples

122

62. Nickel(II) acetylacetonate (Pentane-2,4-dione, nickel(II) derivative)

CAS Registry Number	3264-82-2
CAS Name	Nickel, bis(3-oxobutanoato-O^1,O^3)-
Molecular Formula	

Molecular Weight	256.91, 274.94 (hydrate)
Boiling Point	Not available.
Melting Point	230°C (dec.), 285° (dec.) (hydrate)
Density	Not available.
Refractive Index	Not available.
Safety and Handling	Cancer suspect agent. Hygroscopic.
Reactions	Homogeneous catalyst. Coupling catalyst. Review: *Adv. Organomet. Chem.*, 1979, **17**, 195.
Availability	Aldrich: anhydrous, 95%, £; anhydrous, tech. grade, 90%, p; hydrate, p.
	Lancaster Synthesis: hydrate, p, bulk prices available.

Carbenoid insertion catalyst

A. K. Chakraborti, J. K. Ray, K. K. Kundu, S. Chakrabarty, D. Mukherjee, U. R. Ghatak*, *J. Chem. Soc., Perkin Trans. 1*, 1984, (2), 261-273

Ni(acac)$_2$, cyclohexane
reflux, hv, 7 h

71 - 75 %
45:55 - 90:10 mixtures of isomers
3 examples

Regioselective metal-catalyzed addition of β-dicarbonyl compounds to benzoyl cyanide

M. Basato*, B. Corain*, M. Cofler, A. C. Veronese, G. Zanotti, *J. Chem. Soc., Chem. Commun.*, 1984, (*23*), 1593-1594

Ni(acac)$_2$, ClCH$_2$CH$_2$Cl
reflux, 2.5 h

R^1, R^2= Me, Ph

73 - 87 %
3 examples

Ultrasound preparation of organozinc reagents and their nickel-catalyzed reactions with α,β-unsaturated carbonyl compounds

C. Petrier, J. C. de Souza Barbosa, C. Dupuy, J.-L. Luche*, *J. Org. Chem.*, 1985, **50**(*26*), 5761-5765

Me(CH$_2$)$_6$Br
ultrasound, ZnBr$_2$, Li
PhMe, THF, 0°, 30 min

[Me(CH$_2$)$_6$]$_2$Zn

2-cyclohexenone
Ni(acac)$_2$, 50°, 5 min

(CH$_2$)$_6$Me

88 %
17 examples

63. Nickel(II) chloride

CAS Registry Number	7718-54-9, 7791-20-0 (hexahydrate)
CAS Name	Nickel chloride ($NiCl_2$)
Molecular Formula	$NiCl_2$
Molecular Weight	129.62, 237.71 (hexahydrate)
Boiling Point	973°C (subl.)
Melting Point	1001°C
Density	3.55 kg/m^3
Refractive Index	1.57 (hexahydrate)
Safety and Handling	Cancer suspect agent. Toxic. Deliquescent.
Reactions	Condensation. Reduction.
Availability	Aldrich: hexahydrate, 99.9999%, ££; hexahydrate, 99%, p.
	Johnson Matthey: hexahydrate, crystalline, Specpure®, p; hexahydrate AR, 98%, p; hexahydrate, 99%, p.
	Sigma: hexahydrate, crystalline, p.

Nickel chloride-catalyzed reaction of allyl acetates with dialkyl phosphonates

Lu Xi-Yan*, Z. Jing-Yang, *Huaxue Xuebao*, 1987, **45**(*3*), 312-313

$R^1 = H$, alkyl
$R^2, R^3 =$ alkoxy, Ph
$X = OAc, O_2CCF_3$

55 - 95 %
9 examples

Cross-condensation of cycloalkanones with aldehydes and primary alcohols

T. Nakano, S. Irifune, S. Umano, A. Inada, Y. Ishii*, M. Ogawa, *J. Org. Chem.*, 1987, **52**(*11*), 2239-2244

5 - 48 %
examples

$R^2 = $ alkyl
$n = 1, 2$

$R^1 = H$, alkyl
32 - 80 %
7 examples

$R^1 = Ph$, furyl
-CH=CHPh
67 - 86 %
6 examples

Ultrasonically improved reduction of α,β-unsaturated carbonyl compounds with an aqueous zinc-nickel(II) chloride system

C. Petrier, J.-L. Luche, *Tetrahedron Lett.*, 1987, **28**(*21*), 2347-2350

ultrasound, Zn, NiCl$_2$
MeOCH$_2$CH$_2$OH, H$_2$O
1 - 25 h

$R = H$, alkyl, OEt
$R^1, R^2 = H$, alkyl, aryl

25 - 100 %
10 examples

64. Osmium(VIII) oxide
(Osmic acid, osmium tetroxide)

CAS Registry Number	20816-12-0
CAS Name	Osmium oxide (OsO_4)
Molecular Formula	OsO_4
Molecular Weight	254.20
Boiling Point	130°C
Melting Point	39.5-41°C
Density	4.900 kg/m^3
Refractive Index	Not available.
Safety and Handling	Highly toxic by inhalation, in contact with skin and if swallowed. Corrosive. Oxidizer. Pyrophoric.
Reactions	Oxidation. Reagent for cis-dihydroxylation of double bonds. Reviews: *Synthesis*, 1974, 229; *Chem. Rev.*, 1980, **80**, 187.
Availability	Aldrich: 99.8% (ampoules), £££; 4 wt% in H_2O (ampoules), ££; 2.5 wt% in Bu^{i}OH (stabilized with Bu^{t}OOH), ££. Lancaster Synthesis: 99.8+%, £££ (reagents and literature references are given for regeneration of OsO_4). Johnson Matthey: 2-4% aq. soln (ampoules), p; 0.1-1.0 g units (ampoules), £££. Sigma: (ampoules), £££.

Homologation of carbonyl compounds to α-hydroxy carboxylic esters by diethyl(trimethylsilylethoxymethyl)phosphonate

J. Binder, E. Zbiral*, *Tetrahedron Lett.*, 1986, **27**(*48*), 5829-5832

R^1= alkyl, aryl
R^2= H
or R^1R^2= cycloalkyl

40 - 88 %

25 - 85 %
6 examples

Asymmetric dihydroxylation of alkenes via ligand-accelerated catalysis

E. N. Jacobsen, I. Marko, W. S. Mungall, G. Schroder, K. B. Sharpless*, *J. Am. Chem. Soc.*, 1988, **110**(*6*), 1968-1970

i) , Me_2CO, H_2O
dihydroquinidine p-chlorobenzoate
ii) OsO_4, PhMe, 4°, 17 h
iii) $Na_2S_2O_5$, r.t., 1 h
iv) Na_2SO_4, CH_2Cl_2, 30 min

(+) - threo-, 99 % e.e.
70 %

Stereocontrolled preparation of phospho-sugars from phospholenes

M. Yamashita*, M. Uchimura, A. Iida, L. Parkanayr, J. Clardy, *J. Chem. Soc., Chem. Commun.*, 1988, (*9*), 569-570

OsO_4, $KClO_3$, THF,
H_2O, 45 - 50°, 18 h

+ 2 isomers, 13:1:1 mixture

91 % total yield
3 examples

65. Ozone

CAS Registry Number	10028-15-6
CAS Name	Ozone
Molecular Formula	O_3
Molecular Weight	47.9982
Boiling Point	$-111^\circ C$
Melting Point	$-192.7^\circ C$
Density	2.144 (0°, gas)
Refractive Index	1.2226 (liq.)
Safety and Handling	Toxic by inhalation. Irritating to respiratory system. Highly explosive. Forms explosive peroxides with alkanes, arenes, reacts explosively with N_2O_4, HBr, C_2H_4, N_2, NO.
Reactions	Oxidation. Ether cleavage. Hydroxylation. Review: *Chem. Rev.*, 1958, **58**, 925-1010.
Availability	Not commercially available.
Preparation	Prepared from oxygen using commercially available ozonizer.

Synthesis of heterocyclic fused isoquinolines through N-acyliminium ion intermediates

S. Kano*, Y. Yuasa, S. Shibuya, *Synth. Commun.*, 1985, **15**(*10*), 883-889

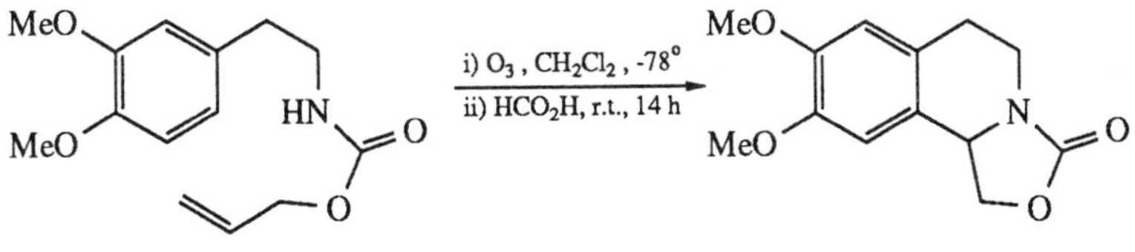

i) O₃, CH₂Cl₂, -78°
ii) HCO₂H, r.t., 14 h

75 %
12 examples

Mild deprotection of benzyl ether protective groups with ozone

P. Angibeaud, J. Defaye*, A. Gadelle, J.-P. Utille, *Synthesis*, 1985, (*12*), 1123-1125

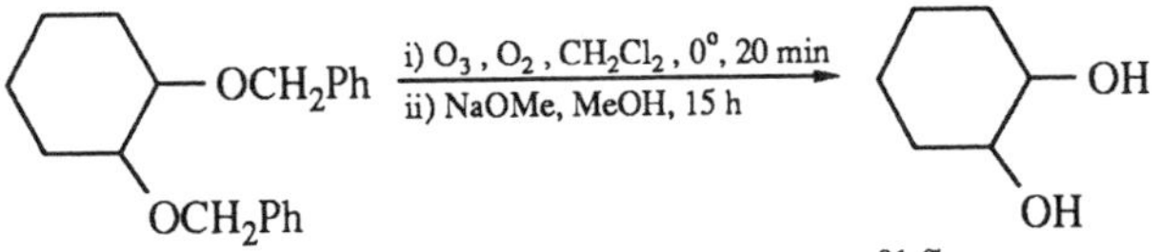

i) O₃, O₂, CH₂Cl₂, 0°, 20 min
ii) NaOMe, MeOH, 15 h

81 %
6 examples

Synthesis of 1,2-benzoquinones from 1,4-benzodioxins

C. Kashima*, A. Tomotake, Y. Omote, *Heterocycles*, 1987, **26**(2), 363-366

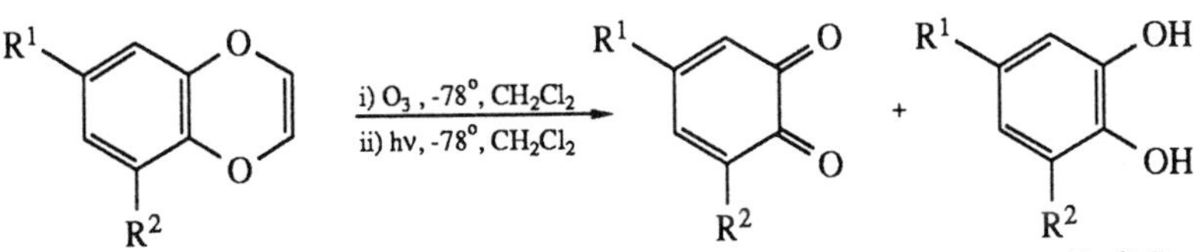

i) O₃, -78°, CH₂Cl₂
ii) hv, -78°, CH₂Cl₂

R¹= H, Cl, alkyl
R²= H, alkyl

17 - 41 %   46 - 51 %
4 examples

66. Palladium(II) acetate

CAS Registry Number	3375-31-3
CAS Name	Acetic acid, compounds, palladium(2+) salt
Molecular Formula	$(AcO)_2Pd$
Molecular Weight	224.49
Boiling Point	Not available.
Melting Point	Not available.
Density	Not available.
Refractive Index	Not available.
Safety and Handling	Harmful.
Reactions	Homogeneous catalyst: *Tetrahedron*, 1981, **37**, 1213. Heck reaction catalyst. Reviews: *Org. React.*, 1982, **27**, 345; *Pure Appl. Chem.*, 1978, **50**, 691; *Acc. Chem. Res.*, 1979, **12**, 146; *Synthesis*, 1970, 225; 1973, 524; 1985, 253; *Tetrahedron*, 1977, **33**, 2615.
Availability	Aldrich: 98%, £££.
	Lancaster Synthesis: £££.
	Sigma: £££.

Palladium-catalyzed intermolecular allylic arylation under exceptionally mild conditions

R. C. Larock*, B. E. Baker, *Tetrahedron Lett.*, 1988, **29**(8), 905-908

PhI + [cyclopentene] →(KOAc, Pd(OAc)$_2$, Bu$_4$NCl / DMF, 25°, 2 days)→ Ph—[cyclopentene]

89 %
11 examples

Multi-step catalysis for the oxidation of alkenes to ketones by molecular oxygen

J.-E. Backvall*, R. B. Hopkins, *Tetrahedron Lett.*, 1988, **29**(23), 2885-2888

R—CH=CH$_2$ →(1/2 O$_2$, Pd(OAc)$_2$, hydroquinone / Fe phthalocyanine, HClO$_4$ / H$_2$O, DMF, 2 - 8 h)→ R—C(O)—CH$_3$

R = alkyl

47 - 85 %
5 examples

Palladium-catalyzed synthesis of conjugated trienes

A. Kasahara*, T. Izumi, N. Kudou, *Synthesis*, 1988, (9), 704-705

Cl—C(O)—CH=CH—C(O)—Cl + CH$_2$=C(R^1)(R^2) →(Pd(OAc)$_2$,EtN O / p-xylene, reflux, 8 h)→ R^2—C(R^1)=CH—CH=CH—CH=C(R^1)(R^2)

44 - 56 %
3 examples

Ph—CH=CH—CH=CH—C(O)—Cl + CH$_2$=CH—R^3 → Ph—CH=CH—CH=CH—CH=CH—R^3

E,E,E 29 - 55 % + Z,E,E 16 - 21 %
4 examples

R^1= H, Me
R^2= Ph, CO$_2$Me, CO$_2$Et
R^3= CN, COMe, Ph, CO$_2$Et

67. Phenylselenenyl chloride

CAS Registry Number	5707-04-0
CAS Name	Benzeneselenenyl chloride
Molecular Formula	PhSeCl
Molecular Weight	191.52
Boiling Point	120°C/20 mmHg
Melting Point	63-65°C
Density	Not available.
Refractive Index	Not available.
Safety and Handling	Highly toxic. Corrosive. Air sensitive.
Reactions	Cyclization. Phenylselenylation. Reviews: *Tetrahedron*, 1978, **34**, 1049; 1985, **41**(*21*), 427-489; *Acc. Chem. Res.*, 1979, **12**, 22; 1984, **17**, 28; S. V. Ley, *Chem. Ind. (London)*, 1985, (*4*), 101-106.
Availability	Aldrich: 98%, ££.
	Lancaster Synthesis: 98%, ££, bulk prices available.

Intramolecular amidoselenation of unsaturated amides leading to lactams

A. Toshimitsu*, K. Terao, S. Uemura, *Tetrahedron Lett.*, 1984, **25**(*51*), 5917-5920

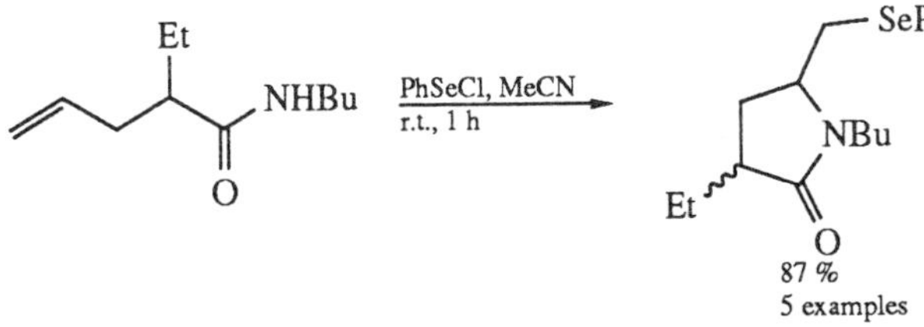

87 %
5 examples

Preparation of chloroalkenes via selenoxide elimination

L. Engman, *Tetrahedron Lett.*, 1987, **28**(*13*), 1463-1466

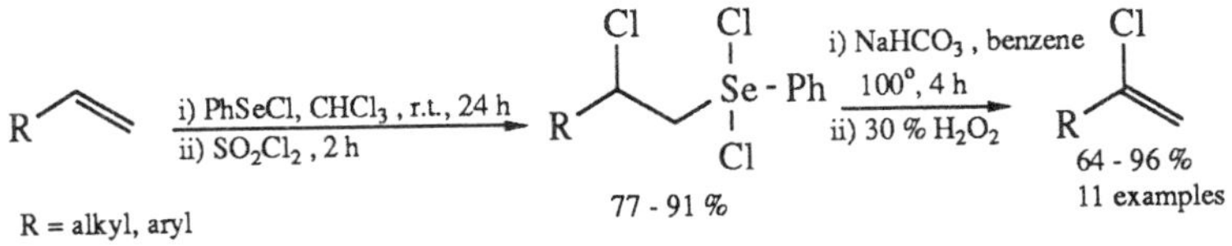

R = alkyl, aryl

77 - 91 %

64 - 96 %
11 examples

Reactions of styrenes and vinyl halides with phenylselenium chloride. Synthesis of chlorophenylethyl ethers, dimethoxyphenylalkanes and α-alkoxy acetals

M. Tiecco*, L. Testaferri*, M. Tingoli, D. Chianelli, D. Bartoli, *Tetrahedron*, 1988, **44**(*8*), 2261-2272, 2273-2282

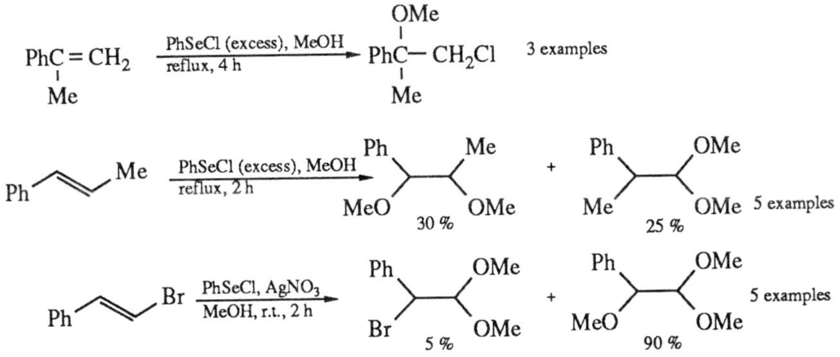

3 examples

5 examples

5 examples

68. *N*-(Phenylseleno)phthalimide

CAS Registry Number	71098-88-9
CAS Name	1*H*-Isoindole-1,3(2*H*)-dione, 2-(phenylseleno)-
Molecular Formula	

Molecular Weight	302.19
Boiling Point	Not available.
Melting Point	181-184°C
Density	Not available.
Refractive Index	Not available.
Safety and Handling	Highly toxic. Moisture sensitive. Air sensitive.
Reactions	Phenylselenylation. Reviews: *Tetrahedron*, 1978, **34**, 1049; 1985, **41**(*21*) 427-489; *Acc. Chem. Res.*, 1979, **12**, 22; 1984, **17**, 28; S. V. Ley, *Chem. Ind. (London)*, 1985, (*4*), 101-106.
Availability	Aldrich: £££. Lancaster Synthesis: £££, bulk prices available.

Synthesis of spiroacetals from alkenyl hydroxy ketones

A. M. Doherty, S. V. Ley*, B. Lygo, D. J. Williams, *J. Chem. Soc., Perkin Trans. 1*, 1984, (*6*), 1371-1377

CSA, MeOH
50°, 30 min

90 %

N-phenylselenophthalimide,
$ZnBr_2$, CH_2Cl_2, r.t., 1-2 h

81 %   SePh
5 examples

THP

Cyanamidoselenenylation of alkenes

R. Hernandez, G. I. Leon, J. A. Salazar*, E. Suarez, *J. Chem. Soc., Chem. Commun.*, 1987, (*4*), 312-314

N-phenylselenophthalimide, H_2NCN
p-TsOH, CH_2Cl_2, 25°, 24 h

NHCN

78 %   SePh
9 examples

Cyclization of hexadienols to tetrahydrofurans with *N*-phenylselenophthalimide

E. Magnol, J. Gore, M. Malacria*, *Bull. Soc. Chim. Fr.*, 1987, (*3*), 455-461

N-phenylselenophthalimide, CH_2Cl_2
-78° to r.t., 3 h

HO

SePh

40 %
7 examples

69. Phosgene
(Carbonyl chloride, carbon oxychloride)

CAS Registry Number 75-44-5

CAS Name Carbonic dichloride

Molecular Formula Cl_2CO

Molecular Weight 98.92

Boiling Point 7.6°C

Melting Point -104°C

Density 1.392 kg/m^3

Refractive Index Not available.

Safety and Handling Very toxic by inhalation.

Reactions Chloroformylation. Carbonylation.

Availability Not commercially available.

Preparation E. E. Hardy, in *Kirk-Othmer Encycl. Chem. Technol.*, 2nd edn, Vol. Suppl., ed. A. Standen, Interscience, NY, pp. 674-683.

Chloroformylation of benzyltetrahydroisoquinolines

J. F. Stambach*, L. Jung, *Tetrahedron*, 1985, **41**(*1*), 169-172

R^1- R^3= H, alkoxy

Cl$_2$CO, Et$_3$N, CHCl$_3$

71 - 79 %
4 examples

Stereo- and regiospecific oxyamination of alkenes via the corresponding bromohydrins

J. Das, *Synth. Commun.*, 1988, **18**(*9*), 907-915

i) Cl$_2$CO , PhMe, Et$_3$N, 0°
ii) CH$_2$Cl$_2$, BnNH$_2$, 0°
iii) NaH, THF, 0°
iv) Li-NH$_3$, THF, -78°

84 %
6 examples

Improved, one-pot synthesis of thiadiazinone dioxides

M. E. Thompson, *Synthesis*, 1988, (*9*), 733-735

R^1CH$_2$SO$_2$NHR^2 $\xrightarrow{\text{BuLi, THF, -78°}}$ $\begin{bmatrix} \text{Li} & \text{Li} \\ | & | \\ R^1\text{CHSO}_2\text{NR}^2 \end{bmatrix}$

R^1= H, Me, Ph
R^2= i-Pr, t-Bu, cyclohexyl, CH$_2$Ph
R^3= i-Pr, t-Bu, Ph, ClC$_6$H$_4$

i) R^{3}CN, -78°- r.t.
ii) Cl$_2$CO or
PhOCOCl, -78°

29 - 45 %
6 examples

i) R^{3}CHO, -78°- r.t.
ii) Cl$_2$CO or
PhOCOCl, -78°

27 - 36 %
2 examples

70. Potassium superoxide

CAS Registry Number 12030-88-5

CAS Name Potassium dioxide

Molecular Formula KO_2

Molecular Weight 71.10

Boiling Point Not available.

Melting Point $380^{\circ}C$

Density 2.14 kg/m^3

Refractive Index Not available.

Safety and Handling Stable when pure. Oxidizer. Reacts violently with Se_2Cl_2, explosively with hydrocarbons. Oxidises As, Sb, Cu, Sn, Zn with incandescence.

Reactions Oxidation.

Availability Aldrich: powder, p; chunks (5-10mm particle size), p.

Sigma: p.

Facile conversion of tosylhydrazones to carbonyl compounds

Y. H. Kim*, H. K. Lee, H. S. Chang, *Tetrahedron Lett.*, 1987, **28**(*37*), 4285-4288

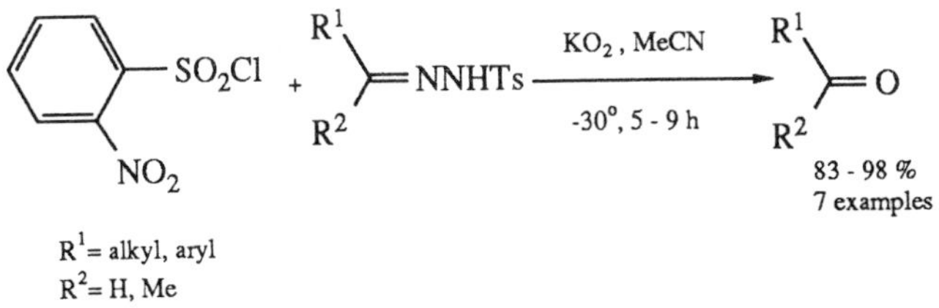

R^1= alkyl, aryl
R^2= H, Me

Oxidation of sulphoxides to sulphones with a peroxysulphur generated *in situ* from nitrobenzenesulphonyl chloride and superoxide

Y. H. Kim*, H. K. Lee, *Chem. Lett.*, 1987, (*8*), 1499-1502

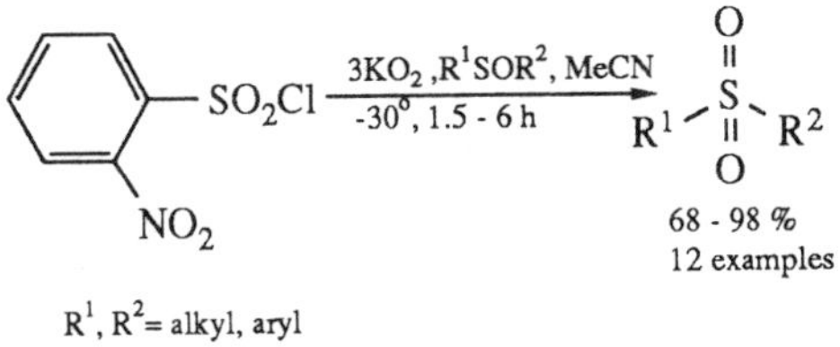

R^1, R^2= alkyl, aryl

Oxidation of arenes to arene oxides by a nitrobenzene peroxysulphur intermediate from nitrobenzenesulphonyl chloride and superoxide

H. H. Lee, K. S. Kim, J. C. Kim, Y. H. Kim*, *Chem. Lett.*, 1988, (*4*), 561-564

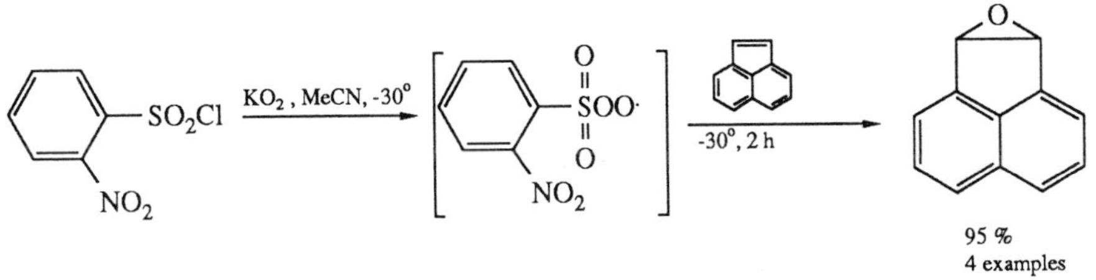

71. Pyridinium chlorochromate (PCC)

CAS Registry Number	26299-14-9
CAS Name	Chromate (1-), chlorotrioxo, (T-4)-, hydrogen, compound with pyridine (1:1)
Molecular Formula	

$$N^+ \quad ClCrO_3^-$$
$$H$$

Molecular Weight	215.56
Boiling Point	Not available.
Melting Point	$205^{o}C$ (dec.)
Density	Not available.
Refractive Index	Not available.
Safety and Handling	Cancer suspect agent. Oxidizer.
Reactions	Stable versatile oxidizing agent. Review: *Synthesis* 1982, 245.
Availability	Aldrich: 98%, p.
	Lancaster Synthesis: 98%, p, bulk prices available.
	Sigma: p.

Facile and selective oxidative cleavage of enol ethers by pyridinium chlorochromate

S. Baskaran, I. Islam, M. Raghavan, S. Chandrasekaran*, *Chem. Lett.*, 1987, (6), 1175-1178

PCC, CH_2Cl_2
$25°$, 1 h

85 %
11 examples

Allyl oxidation of Δ^5-steroids with pyridinium chlorochromate

E. J. Parish*, T.-Y. Wei, *Synth. Commun.*, 1987, **17**(*10*), 1227-1233

PCC, C_6H_6 , reflux
or PCC, DMSO, $100°$
or PDC, pyridine $100°$

BzO

64 - 89 %
5 examples

Pyridinium chlorochromate oxidation of aldehydes to carbamoyl/acyl azides or carboxylic acids

P. S. Reddy, P. Yadagiri, S. Lumin, D.-S. Shin, J. R. Falck*, *Synth. Commun.*, 1988, **18**(*5*), 545-551

PCC, NaN_3 , CH_2Cl_2
$45°$, 2 - 3 h
R = alkyl, aryl

RCHO

$RNHCON_3$ + $RCON_3$

10 - 65 %
26:74 - 100:1
9 examples

PCC, NaCN, THF
$45°$, 2 - 3 h

R = alkyl

RCO_2H

30 - 78 %
5 examples

72. Pyridinium dichromate (PDC)

CAS Registry Number	20039-37-6
CAS Name	Chromic acid, compound with pyridine (1:2)
Molecular Formula	

$$\left[\begin{array}{c} \bigcirc \\ N^+ \\ | \\ H \end{array} \right]_2 Cr_2O_7{}^{2-}$$

Molecular Weight	376.21
Boiling Point	Not available.
Melting Point	152-153°C
Density	Not available.
Refractive Index	Not available.
Safety and Handling	Cancer suspect agent. Oxidizer.
Reactions	Oxidizing agent complementary to PCC for ROH with acid-sensitive groups: *Tetrahedron Lett.*, 1979, **20**, 399; 1980, **21**, 731.
Availability	Aldrich: 98%, p.

Conversion of aldehydes to methyl esters using pyridinium dichromate

B. O'Connor, G. Just*, *Tetrahedron Lett.*, 1987, **28**(*28*), 3235-3236

$$R\text{-}CHO \xrightarrow[\text{N}_2\text{, r.t., 20 h}]{\text{PDC, MeOH, DMF}} R\text{-}COOMe$$

R = alkyl, cycloalkyl, steroid

60 - 87 %
4 examples

Allylic and benzylic oxidations with *tert*-butyl hydroperoxide-pyridinium dichromate

N. Chidambaram, S. Chandrasekaran*, *J. Org. Chem.*, 1987, **52**(*22*), 5048-5057

t-BuOOH, PDC, benzene
10°, 15 min; 25°, 11 h

40 % yield
69 % conversion
12 examples

Synthesis of ω-alkynyl aldehydes and ketones via oxidation of ω-alkynyl alcohols with pyridinium dichromate

D. E. Bierer, G. W. Kabalka*, *Org. Prep Proced. Int.*, 1988, **20**(*1-2*), 63-72

$$RC{\equiv}CCH_2(CH_2)_nCH(OH)R' \xrightarrow[\text{r.t., 20 h}]{\text{PDC, CH}_2\text{Cl}_2} RC{\equiv}CCH_2(CH_2)_nCR'$$

R = H, SiMe$_3$
R'= H, Me
n = 1 - 7

21 - 84 %
8 examples

144

73. Pyridinium *p*-toluenesulphonate (PPTS)

CAS Registry Number   24057-28-1

CAS Name   Pyridine, 4-methylbenzenesulfonate

Molecular Formula

Molecular Weight   251.31

Boiling Point   Not available.

Melting Point   117-119°C

Density   Not available.

Refractive Index   Not available.

Safety and Handling   Moisture sensitive. Irritant.

Reactions   Efficient catalyst for THP ether preparation (OH protection): *J. Org. Chem.*, 1977, **42**, 3772. Cleavage of MEM and MOM ethers: *Synth. Commun.*, 1983, 1021.

Availability   Aldrich: 98%, £.

Lancaster Synthesis: 98%, £, bulk prices available.

A mild and stereospecific conversion of vicinal diols into alkenes

M. Ando*, H. Ohhara, K. Takase, *Chem. Lett.*, 1986, (6), 879-882

i) PPTS, CH(OMe)$_3$, CH$_2$Cl$_2$
ii) sealed tube, 140°, 4 h

95 %
19 examples

Synthesis of acyl and keto cyclic ethers

T. Satoh, K. Iwamoto, K. Yamakawa*, *Tetrahedron Lett.*, 1987, **28**(*23*), 2603-2606

PPTS, EtOH or PrOH
50°- reflux, 0.5 - 48 h

R' = alkyl
n = 1 - 11

52- 90 %
7 examples

n = 3, 4

PPTS, EtOH or PrOH
reflux, 4 h - 3 days

11 - 60 %
2 examples

Selective monobenzylidenation of monosaccharides with α,α-dimethoxytoluene

J. J. Patroni, R. V. Stick*, B. W. Skelton, A. H. White, *Aust. J. Chem.*, 1988, **41**(*1*), 91-102

PhCH(OMe)$_2$, PPTS
DMF, 100°, 2.5 h

19 %
9 examples

146

74. Raney® nickel

CAS Registry Number   7440-02-0

CAS Name   See Nickel, uses and miscellaneous, catalysts

Molecular Formula   Not available.

Molecular Weight   Not available.

Boiling Point   Not available.

Melting Point   Not available.

Density   Not available.

Refractive Index   Not available.

Safety and Handling   Cancer suspect agent. Flammable solid.

Reactions   Hydrogenation catalyst.

Availability   Aldrich: active catalyst (50% slurry in H_2O, pH10), p.

Sigma: active catalyst (50% slurry in H_2O, pH9), p.

Stereoselective synthesis of *erythro*- and *threo*-1,2-diols from diketo sulphides via *cis*-3,4-dihydroxy thiolanes

J. Nakayama*, S. Yamaoka, M. Hoshino, *Tetrahedron Lett.*, 1987, 28(*16*), 1799-1802

Ph $\quad$ O O $\quad$ Ph
TiCl$_4$-Zn, THF, 0°

Ph $\quad$ Ph
HO $\quad$ OH
S
82 %

Raney Ni, EtOH, reflux

Ph $\quad$ Ph
HO $\quad$ OH
Me $\quad$ Me
75 %
10 examples

Reaction of Raney nickel with primary, secondary and tertiary alcohols

M. E. Krafft*, W. J. Crooks III, B. Zorc, S. E. Milczanowski, *J. Org. Chem.*, 1988, 53(*14*), 3158-3163

MeO $\quad$ (CH$_2$)$_9$ $\quad$ OH

Raney Ni, Δ, PhMe
3.5 h

MeO $\quad$ (CH$_2$)$_9$CH$_3$

73 %
8 examples

OH

Raney Ni, Δ, benzene
9 h

93 %
10 examples

OH

i) Raney Ni, Δ, PhMe
40 min
ii) EtOH, EtOAc, H$_2$
1 atm., 8 h, r.t., Pd(C)

99 %
12 examples

Condensation of nitro compounds with alcohols catalyzed by Raney nickel

A. A. Banerjee, D. Mukesh*, *J. Chem. Soc., Chem. Commun.*, 1988, (*18*), 1275-1276

H
N
NO$_2$

i-PrOH, Raney Ni, 60°

H
N
NHPr-i
95 %
5 examples

148

75. Rhodium(II) acetate

CAS Registry Number   5503-41-3

CAS Name   Acetic acid, compounds, rhodium(2+) salt

Molecular Formula $[(AcO)_2Rh]_2$

Molecular Weight   441.99

Boiling Point   Not available.

Melting Point   Not available.

Density   Not available.

Refractive Index   Not available.

Safety and Handling   Not available.

Reactions   Catalyst for formation of carbenes from diazo compounds. Homogeneous catalyst: *Tetrahedron Lett.*, 1980, **21**, 4039.

Availability   Aldrich: dimer, £££.

Synthesis of γ,δ-unsaturated carbonyl compounds from allyl sulphides and α-diazo carbonyl compounds

S. Takano*, S. Tomita, M. Takahashi, K. Ogasawara, *Chem. Lett.*, 1987, (*8*), 1569-1570

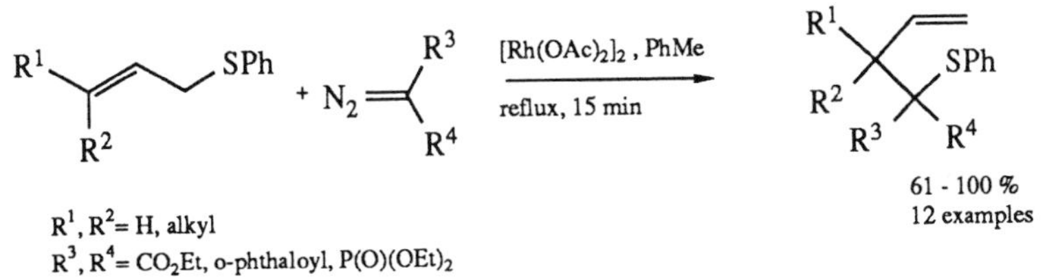

R1
SPh
+ N2
R3
R4
R2
[Rh(OAc)2]2 , PhMe
reflux, 15 min
61 - 100 %
12 examples
R1, R2= H, alkyl
R3, R4= CO2Et, o-phthaloyl, P(O)(OEt)2

Preparation and rhodium(II) acetate catalyzed cyclization of ω-hydroxy-, -mercapto-, and -amino-α-diazo-β-keto esters

C. J. Moody*, R. J. Taylor, *Tetrahedron Lett.*, 1987, **28**(*44*), 5351-5352

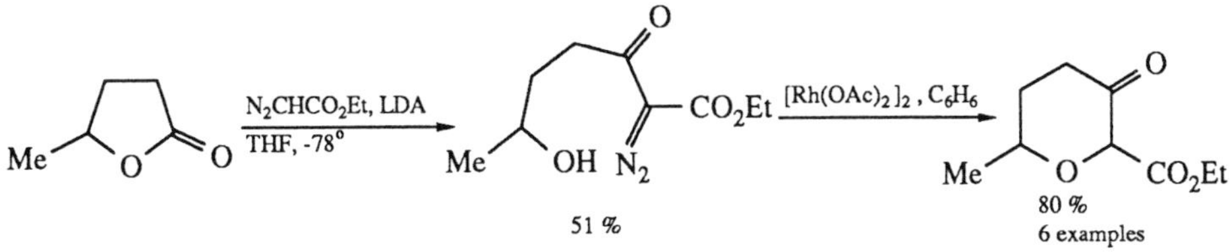

Me
O
N2CHCO2Et, LDA
THF, -78°
OH
N2
CO2Et
51 %
[Rh(OAc)2]2 , C6H6
80 %
6 examples

Synthesis of tosylpyrrolidinones via ketenes and carbenes

A. Saba*, A. Selva, *Heterocycles*, 1987, **27**(*4*), 867-870

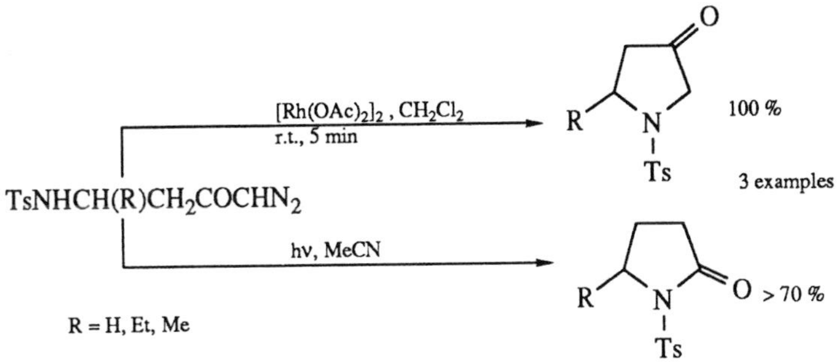

TsNHCH(R)CH2COCHN2
[Rh(OAc)2]2 , CH2Cl2
r.t., 5 min
100 %
3 examples
hv, MeCN
> 70 %
R = H, Et, Me
N
Ts

76. Rhodium(III) chloride

CAS Registry Number   10049-07-7, 20765-98-4 (hydrate)

CAS Name   Rhodium chloride ($RhCl_3$)

Molecular Formula $RhCl_3$

Molecular Weight   209.26, 263.31 (trihydrate)

Boiling Point $800^{\circ}C$ (subl.)

Melting Point $450\text{-}500^{\circ}C$ (dec.)

Density   Not available.

Refractive Index   Not available.

Safety and Handling   Highly toxic. Hygroscopic.

Reactions   Condensation catalyst. Hydrogenation catalyst.

Availability   Aldrich: anhydrous, £££; hydrate, £££.

Lancaster Synthesis: hydrate, £££, bulk prices available.

Johnson Matthey: hydrate, crystalline, £££.

Synthesis of methyl 2-arylpropionates from methyl arylacetates and formaldehyde

K. Takeuchi, Y. Sugi, T. Matsuzaki, H. Arakawa, K. Bando, *Chem. Ind.* (*London*), 1985, (*13*), 446-447

$$ArCH_2CO_2Me \ + \ HCHO \ + \ CO \ \xrightarrow[H_2O, MeOH]{RhCl_3, N\text{-methylmorpholine,}} ArCH(Me)CO_2Me$$

54 - 83 %
6 examples

Selective C-C bond hydrogenation of unsaturated nitro compounds in the presence of a rhodium(III) chloride-Aliquat 336 system

I. Amer, T. Bravdo, J. Blum*, K. P. C. Vollhardt, *Tetrahedron Lett.*, 1987, **28**(*12*), 1321-1322

$$3\text{-}O_2NC_6H_4CH{=}CH_2 \ \xrightarrow[MeNO_2, H_2O, 7\,h]{H_2, RhCl_3, Aliquat\ 336,} 3\text{-}O_2NC_6H_4Et$$

70 % conversion
5 examples

Synthesis of β-siloxy esters by condensation of carbonyl compounds and trimethylsilane with α,β-unsaturated esters catalyzed by rhodium(III) chloride

A. Revis*, T. K. Hilty, *Tetrahedron Lett.*, 1987, **28**(*41*), 4809-4812

95 %
8 examples

77. Ruthenium(III) chloride

CAS Registry Number	10049-08-8, 14898-67-0 (hydrate)
CAS Name	Ruthenium chloride (RuCl$_3$)
Molecular Formula	RuCl$_3$
Molecular Weight	207.43, 261.47 (trihydrate)
Boiling Point	Not available.
Melting Point	500°C (dec.)
Density	3.110 kg/m^3
Refractive Index	Not available.
Safety and Handling	Corrosive. Hygroscopic.
Reactions	Catalyst for NaOH induced rearrangment of *sec*-allyl alcohols to saturated ketones.
Availability	Aldrich: £££; hydrate, £££.
	Lancaster Synthesis: hydrate, ££, bulk prices available.
	Johnson Matthey: hydrate, crystalline powder, ££; aq. soln, £££.

Phase transfer-catalyzed allylic oxidation by sodium periodate

C. Singh, *Indian J. Chem., Sect. B*, 1985, **24**(*8*), 859

Reaction conditions: NaIO$_4$, RuCl$_3$, CTAB, CCl$_4$, H$_2$O, r.t., 1 h

>90 %
3 examples

CTAB = cetyltrimethylammonium bromide

Synthesis of dialkylformamides

G. Bitsi, G. Jenner*, *J. Organomet. Chem.*, 1987, **330**(*3*), 429-435

$$\underset{R}{\overset{R}{>}}NH \xrightarrow[200°, 2h]{CO\ (450\ bar),\ RuCl_3,\ MeOH} \underset{R}{\overset{R}{>}}NCHO$$

R = alkyl

5 - 95 % conversion
80 - 89 % selectivity
6 examples

Selective oxidation of alcohols by a hydrogen peroxide-ruthenium chloride system under phase-transfer conditions

G. Barak, J. Dakka, Y. Sasson*, *J. Org. Chem.*, 1988, **53**(*15*), 3553-3555

RCH$_2$OH $\longrightarrow$ RCO$_2$H 85 - 89 % 4 examples

RR'CHOH $\xrightarrow{\text{H}_2\text{O}_2,\ \text{RuCl}_3,\ \text{CH}_2\text{Cl}_2,\ \text{didecyldimethylammonium bromide, 80°}}$ RCOR' 82 - 90 % 3 examples

ArCH$_2$OH $\longrightarrow$ ArCHO 45 - 95 % 4 examples

154

78. Ruthenium(IV) oxide

CAS Registry Number	12036-10-1, 32740-79-7 (hydrate)
CAS Name	Ruthenium oxide (RuO_2)
Molecular Formula	RuO_2
Molecular Weight	133.07
Boiling Point	Not available.
Melting Point	Not available.
Density	6.970 kg/m^3
Refractive Index	Not available.
Safety and Handling	Hygroscopic.
Reactions	Oxidation. In-situ generation of ruthenium tetroxide. Review: *Rev. Pure Appl. Chem.*, 1972, **22**, 47-54.
Availability	Aldrich: 99.9%, £££; hydrate, £££.
	Johnson Matthey: hydrate, powder, £££; anhydrous, £££.

Phase-transfer promoted oxidation of secondary alcohols to ketones

P. E. Morris, Jr., D. E. Kiely*, *J. Org. Chem.*, 1987, **52**(6), 1149-1152

Ph$_3$COH$_2$C, OH, O, O, BnO → RuO$_2$, NaIO$_4$, CHCl$_3$ / PhCH$_2$Et$_3$NCl, 28 h → Ph$_3$COH$_2$C, O, O, O, BnO

97 %
6 examples

Ruthenium oxide oxidation of *N*-acyl alkylamines for imide synthesis

K. Tanaka*, S. Yoshifuji, Y. Nitta, *Chem. Pharm. Bull.*, 1987, **35**(1), 364-369

$$RCH_2NHCOR' \xrightarrow[0°\text{- r.t., 4 - 120 h}]{RuO_2. xH_2O, NaIO_4, AcOEt, H_2O} RCONHCOR'$$

R = alkyl, aryl
R'= alkyl

25 - 96 %
11 examples

Preparation of 1,2-diketones from acetylenes via a mild oxidation method

R. Zibuck, D. Seebach*, *Helv. Chim. Acta*, 1988, **71**(1), 237-240

Et——CH$_2$O$_2$CBu-t $\xrightarrow[\text{MeCN, H}_2\text{O, r.t., 30 min}]{\text{RuO}_2, \text{NaIO}_4, \text{CCl}_4}$ Et, O, O, CH$_2$O$_2$CBu-t

82 %
10 examples

79. Samarium(II) iodide

CAS Registry Number	32248-43-4
CAS Name	Samarium iodide (SmI$_2$)
Molecular Formula	SmI$_2$
Molecular Weight	404.16
Boiling Point	1580°C
Melting Point	527°C
Density	Not available.
Refractive Index	Not available.
Safety and Handling	Not available.
Reactions	Prepared *in situ* from Sm metal and ICH$_2$CH$_2$I in THF: *J. Am. Chem. Soc.*, 1980, **102**(*8*), 2693-2698. Coupling catalyst. Reduction. As electron donor: *J. Chem. Soc. Chem. Commun.*, 1982, (*12*), 709-710.
Availability	Johnson Matthey: ultra dry (ampoule under Ar), 99.99%, £££.

Samarium diiodide induced reductive coupling of α,β-unsaturated esters with carbonyl compounds: γ-lactone synthesis

S. Fukuzawa*, A. Nakanishi, T. Fujinami, S. Sakai, *J. Chem. Soc., Perkin Trans. 1*, 1988, (7), 1669-1675

R^1 = alkyl, aryl
R^2 = H, alkyl, Ph
R^3, R^4 = H, Me

40 - 82 %
17 examples

Samarium diiodide promoted intramolecular pinacolic coupling reactions

G. A. Molander*, C. Kenny, *J. Org. Chem.*, 1988, 53(9), 2132-2134

R = alkyl, Ph
R^1 = alkyl, H
R^2 = OEt, NMe$_2$, NEt$_2$
n = 1, 2

35 - 82 %
3.1 - 200:1 selectivity
9 examples

New synthesis of 1,2-glycol monoethers via samarium diiodide mediated decarbonylation of α-alkoxy acid chlorides

M. Sasaki, J. Collin, H. B. Kagan*, *Tetrahedron Lett.*, 1988, 28(38), 4847-4850

i) SmI$_2$, THF, r.t., 1 min
ii) HCl, H$_2$O

53 %
11 examples

80. Selenium(IV) oxide (Selenium dioxide)

CAS Registry Number	7446-08-4
CAS Name	Selenium oxide (SeO_2)
Molecular Formula	SeO_2
Molecular Weight	110.96
Boiling Point	Not available.
Melting Point	315°C (subl.)
Density	3.950 kg/m^3
Refractive Index	Not available.
Safety and Handling	Highly toxic. Corrosive. Irritant to respiratory system, eyes and skin.
Reactions	Oxidation. Reagent for allylic oxidation of alkenes and acetylenes.
Availability	Aldrich: 99.999%, ££; 99.9+%, p; 99.8%, p; 99%, p.
	Lancaster Synthesis: 99.8%, p, bulk prices available.
	Johnson Matthey: powder, Specpure™, ££; powder, 99%, p.
	Sigma: white to pink crystals, p.

Oxidation of 2,4-alkadienoic esters with selenium dioxide. Synthesis of furans and selenophenes

S. Tsuboi, S. Mimura, S. Ono, K. Watanabe, A. Takeda*, *Bull. Chem. Soc. Jpn.*, 1987, **60**(5), 1807-1812

R, R' = alkyl

SeO$_2$, benzene
PhBr, xylene or
DME, reflux,
0.5 - 26 h

17 - 58 %   2 - 48 %

8 examples

Oxidation of cyclic hemiacetals with selenium dioxide

K. Kanai*, I. Tomoskozi, *Synthesis*, 1988, (7), 544-545

SeO$_2$, dioxane
80°, 1 h

50 %
8 examples

Selenium dioxide-catalyzed conversion of alcohols to alkyl chlorides by chlorotrimethylsilane

J. G. Lee*, K. K. Kang, *J. Org. Chem.*, 1988, **53**(15), 3634-3637

ROH $\xrightarrow[\text{20°- reflux, 1 - 7 h}]{\text{Me}_3\text{SiCl, SeO}_2 \text{ , CCl}_4}$ RCl   60 - 100 %
5 examples

ROH = primary, secondary or tertiary alcohol

160

81. Silica gel

CAS Registry Number 7631-86-9

CAS Name Silica gel

Molecular Formula $(SiO_2)_n$

Molecular Weight 60.09 (SiO_2)

Boiling Point Not available.

Melting Point Not available.

Density Not available.

Refractive Index Not available.

Safety and Handling Hygroscopic.

Reactions Chromatographic applications. Catalyst for cyclization. Drying agent. Inert catalyst support.

Availability Aldrich: Wide variety of grades and mesh/particle sizes for chromatographic applications, many 99+%, bulk price range (1-10kg) ~£10-£100.

Sigma: Variety of types and mesh/particle sizes for chromatographic applications, price range (10g-5kg) ~£5-£250; for desiccation and humidity indicators, price range (250g-5kg) ~£5-£60.

Silica gel assisted reductive cyclization of nitropiperidinostyrenes to indoles

M. Kawase, A. K. Sinhababu, R. T. Borchardt*, *J. Heterocyclic Chem.*, 1987, **24**(*6*), 1499-1501

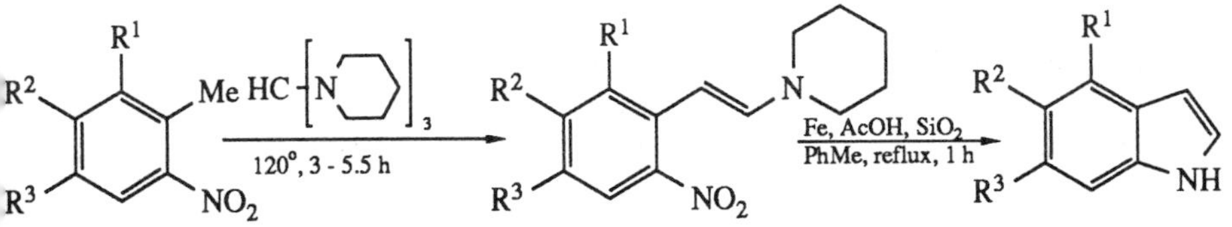

R^1- R^3= H, alkoxy, halo

62 - 94 %
7 examples

Use of moist silica gel for obtaining α-ethylenic carbonyl compounds from β-alkylthio or β-phenylthio allylic alcohols

M. Pellet*, F. Huet, *Tetrahedron*, 1988, **44**(*14*), 4463-4468

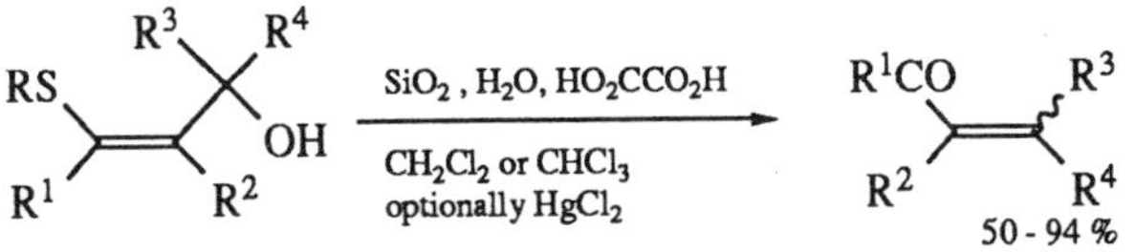

50 - 94 %
7 examples

Silica gel catalyzed cyclizations of mixed ketene acetals

D. Schinzer*, M. Kalesse, J. Kabbara, *Tetrahedron Lett.*, 1988, **29**(*41*), 5241-5244

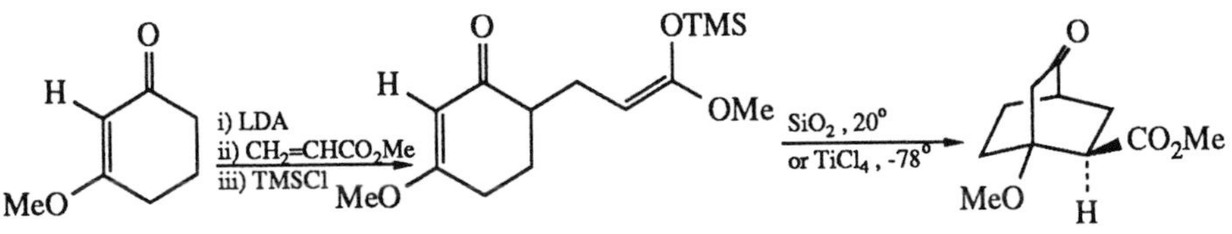

>98 %, 72 % e.e.
6 examples

162

82. Silver trifluoromethanesulphonate (Silver triflate)

CAS Registry Number   2923-28-6

CAS Name   Trifluoromethanesulfonic acid, silver salt

Molecular Formula CF_3SO_3Ag

Molecular Weight   256.94

Boiling Point   Not available.

Melting Point 356°

Density   Not available.

Refractive Index   Not available.

Safety and Handling   Irritant. Light sensitive. Moisture sensitive.

Reactions   Cyclization. Precursor to alkyl methanesulphonates - alkylating agents for aromatic compounds.
Review: *Chem. Rev.*, 1977, **77**, 69.

Availability   Aldrich: 99+%, ££.

Lancaster Synthesis: 99%, ££, bulk prices available.

Synthesis of functionalized carbocycles via silver ion assisted episulphonium ion cyclization

E. Edstrom, T. Livinghouse*, *J. Chem. Soc., Chem. Commun.*, 1986, (4), 279-280

i) PhSCl, CH$_2$Cl$_2$, ClCH$_2$CH$_2$Cl
-78°; -78 to -40°, 1 h
ii) AgOTf, -78°, 1 h; -30 to
-20°, 12 h; 0°, 4 h

79 %
11.4:1 mixture
2 examples

C-Glucosidation of β-keto esters and ketones via enamines

P. Allevi*, M. Anastasia, P. Ciuffreda, A. Fiecchi, A. Scala, *J. Chem. Soc., Chem. Commun.*, 1988, (*1*), 57-58

i) MeC=CHCO$_2$Me

AgOTf, 3A mol. sieves
CH$_2$Cl$_2$, r.t., dark, 10 min
ii) NaCl, H$_2$O, 15 min

85 %
9:1 epimeric mixture
7 examples

Diastereoselective, silver(I)-catalyzed cyclizations of acetylenic isoureas to oxazolidines and oxazines; acetic acid-induced conversion of the alkylideneoxazines to 2-*N*-substituted (1*Z*,3*E*)-dienes

W. Clegg, S. P. Collingwood, B. T. Golding*, S. M. Hodgson, *J. Chem. Soc., Chem. Commun.*, 1988, (*17*), 1175-1176

AgOTf, benzene
r.t. to reflux

n = 1 only
AcOH, CHCl$_3$, Δ

R^1 = alkyl
R^2 = H, Me, CH$_2$OPh
n = 0, 1

7 examples

4 examples

83. Sodium cyanoborohydride

CAS Registry Number	25895-60-7
CAS Name	Borate(1-), (cyano-C)trihydro-, sodium, (T-4)-
Molecular Formula	NaBH$_3$CN
Molecular Weight	62.84, 65.87 (NaBD$_3$CN)
Boiling Point	Not available.
Melting Point	242°C(dec.)
Density	1.083 kg/m^3 (NaOH soln)
Refractive Index	Not available.
Safety and Handling	Highly toxic. Flammable solid. Very hygroscopic.
Reactions	Highly selective reducing agent. Reductive cyclization. Ring cleavage. Reviews: *Synthesis*, 1975, 135; *Org. Prep. Proced. Int.*, 1979, **11**, 201.
Availability	Aldrich: 95%, £; 5M in aq. NaOH (1M), p; 1M in THF, under N$_2$ in Sure/Seal™ bottles, p. Also available: NaBD$_3$CN, £££. Lancaster Synthesis: 95%, £, bulk prices available. Sigma 90-95%, £. Also available: NaBD$_3$CN, £££.

Reductive cyclization of keto esters with sodium cyanoborohydride. Synthesis of γ- and δ-lactones

K. F. Podraza, *J. Heterocyclic Chem.*, 1987, **24**(*1*), 293-295

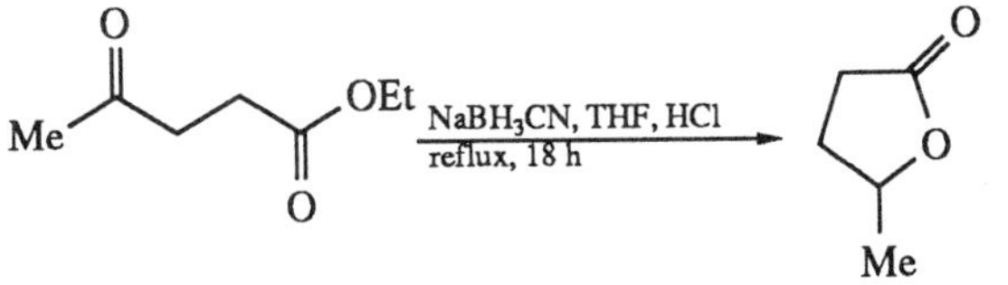

81 %
4 examples

Reductive ring cleavage of cycloadenosines

M. Sako, T. Saito, K. Kameyama, K. Hirota, Y. Maki*, *J. Chem. Soc., Chem. Commun.*, 1987, (*17*), 1298-1299

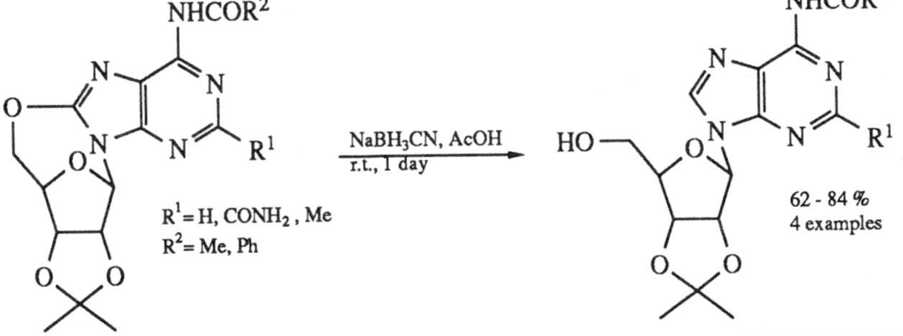

R^1= H, CONH$_2$, Me
R^2= Me, Ph

62 - 84 %
4 examples

Reduction of sugar aldoximes to terminal deoxy hydroxyamino sugars

J. M. J. Tronchet*, G. Zosimo-Landolfo, N. Bizzozero, D. Cabrini, F. Habashi, E. Jean, M. Geoffroy, *J. Carbohydr. Chem.*, 1988, **7**(*1*), 169-186

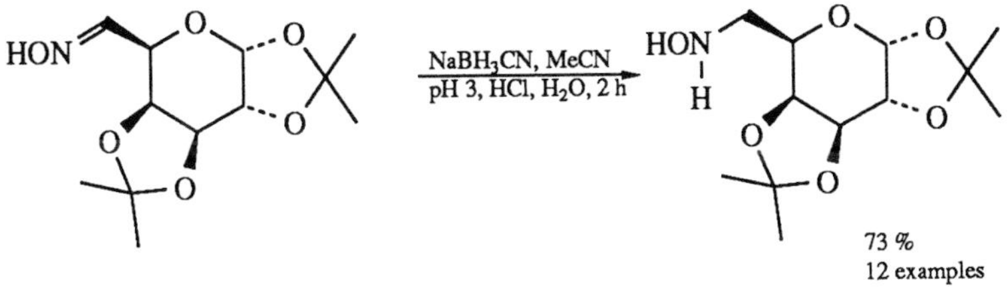

73 %
12 examples

84. Sodium hydrogen telluride

CAS Registry Number   65312-92-7

CAS Name   Sodium hydrogen telluride

Molecular Formula   NaHTe

Molecular Weight   151.60

Boiling Point   Not available.

Melting Point   Not available.

Density   Not available.

Refractive Index   Not available.

Safety and Handling   Not available.

Reactions   Selective reduction. Debromination of vicinal dibromides. Synthesis of [123]Te-labelled radiopharmaceuticals.

Availability   Not commercially available.

Preparation   Prepared by heating to Te with excess $NaBH_4$ in EtOH under N_2, to give wine-coloured solution. D. H. R. Barton, S. W. Crombie, *J. Chem. Soc., Perkin Trans. 1*, 1975, (*16*), 1574.

Pyrazines from α-azido ketones

H. Suzuki*, T. Kawaguchi, T. Takaoka, *Bull. Chem. Soc. Jpn.*, 1986, **59**(2), 665-666

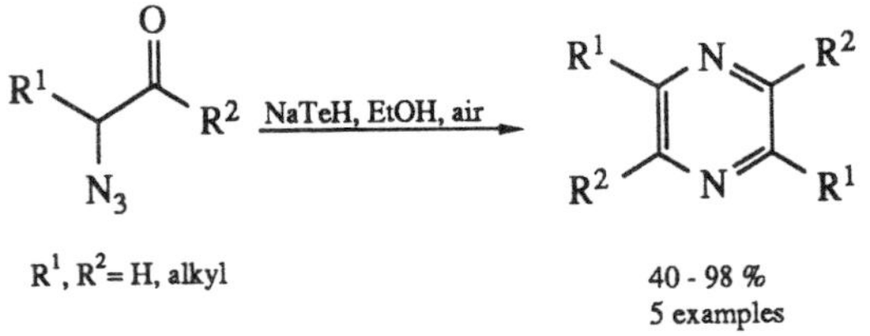

$R^1, R^2 = H$, alkyl

40 - 98 %
5 examples

Synthesis of isopropylmalonates

X. Huang, L. Xie, *Synth. Commun.*, 1986, **16**(*13*), 1701-1707

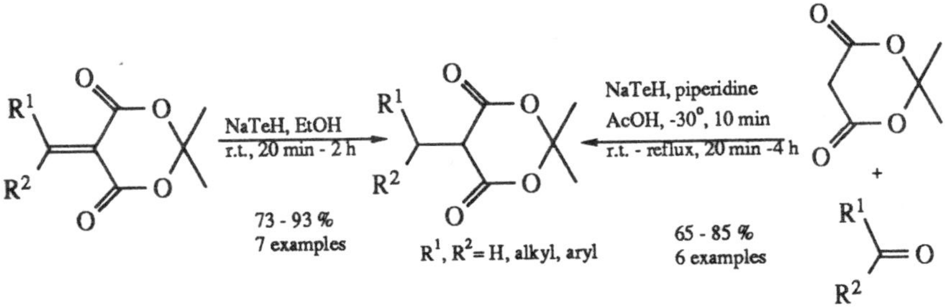

73 - 93 %
7 examples

$R^1, R^2 = H$, alkyl, aryl

65 - 85 %
6 examples

Facile cleavage of haloethyl esters with sodium hydrogen telluride

J. Chen, X.-J. Zhou*, *Synth. Commun.*, 1987, **17**(2), 161-164

$$RCO_2CH_2CH_2X \xrightarrow{\text{NaTeH, EtOH, r.t., 2 - 60 min}} RCO_2H$$

R = alkyl, aryl
X = Br, Cl

84 - 92 %
5 examples

85. Tetrabutylammonium fluoride (TBAF)

CAS Registry Number   429-41-4

CAS Name   Butylaminium, N,N,N-tributyl-, fluoride

Molecular Formula $[Me(CH_2)_3]_4NF$

Molecular Weight   261.47, 315.52 (trihydrate)

Boiling Point   Not available.

Melting Point   62-63°C (trihydrate)

Density   0.903 kg/m^3

Refractive Index   Not available.

Safety and Handling   Irritant. Hygroscopic.

Reactions   Cleavage of silyl ethers. Efficient base. Effective reagent for fluoride-induced aldol condensation of silyl enol ethers with aldehydes: *J. Am. Chem. Soc.*, 1975, **99**, 3257; 1982, **104**, 1025.

Review: *Chem. Rev.*, 1980, **80**, 429.

Availability   Aldrich: hydrate, 99%, £; 1M in THF (5 wt % H$_2$O), p.

Lancaster Synthesis: 1M in THF, p, bulk prices available.

Sigma: 1M in THF; trihydrate; enquire for details.

β-Hydroxyphenylsulphides by fluoride ion induced reaction of phenylthiomethyltrimethylsilane with aldehydes and ketones

J. Kitteringham*, M. B. Mitchell, *Tetrahedron Lett.*, 1988, **29**(*27*), 3319-3322

$$\text{PhSCH}_2\text{SiMe}_3 + \text{RCOR}^1 \xrightarrow[\text{30 min, then r.t., 1.5 h}]{\text{Bu}_4\text{NF, THF, N}_2\text{, 5°}} \text{PhSCH}_2\overset{\displaystyle R}{\underset{\displaystyle R^1}{C}} - \text{OR}^2$$

R = Ph, Me, CH=CHPh
R^1 = H, Me, Ph
R^2 = H, SiMe$_3$

45 - 96 %
7 examples

Effective deprotection of 2-(trimethylsilylethoxy)methyl ethers

T. Kan, M. Haslimoto, M. Yanaguja, H. Shirahama*, *Tetrahedron Lett.*, 1988, **29**(*42*), 5417-5418

$$\text{OSEM} \xrightarrow[\text{r.t., 1 h}]{\text{Bu}_4\text{NF, HMPA, 4A mol.sieves}} \text{OH}$$

94 %
6 examples

SEM = 2-(trimethylsilylethoxy)methyl

Facile, efficient heterogeneous nucleophilic fluorination without solvent

G. Bram*, A. Loupy, P. Pigeon, *Synth. Commun.*, 1988, **18**(*14*), 1661-1667

$$\text{n-C}_8\text{H}_{17}\text{Br} \xrightarrow{\text{Bu}_4\text{NF. 3H}_2\text{O, 20 h, 60°}} \text{n-C}_8\text{H}_{17}\text{F} + \text{C}_8\text{H}_{16}$$

71 %   13 %   30 examples

86. Thionyl chloride (Sulphinyl chloride)

CAS Registry Number	7719-09-7
CAS Name	Thionyl chloride
Molecular Formula	$SOCl_2$
Molecular Weight	118.97
Boiling Point	79°C
Melting Point	-105°C
Density	1.631 kg/m^3
Refractive Index	1.5140
Safety and Handling	Corrosive. Irritating to respiratory system and lachrymatory. Reacts violently with water.
Reactions	Chlorination. Oxidation.
Availability	Aldrich: 99+%, p; 99+% (in poly-coated bottle), p; 97%, p; 97% (in poly-coated bottle), p; 2M in CH_2Cl_2, under N_2 in Sure/Seal™ bottles, p.

Esterification at room temperature in the absence of solvent

B. Kumar*, R. K. Verma, *Synth. Commun.*, 1984, **14**(*14*), 1359-1363

$$RCO_2H \ + \ R'OH \ \xrightarrow[\text{no solvent}]{SOCl_2 \text{ , r.t., overnight}} \ RCO_2R'$$

R = alkyl, aryl
R' = alkyl

43 - 82 %
14 examples

Synthesis of α-phenylthioaldehydes

B. J. M. Hansen, R. M. Peperzak, A. de Groot*, *Recl. Trav. Chim. Pays-Bas*, 1987, **106**(*9*), 489-494

R^1, R^2 = alkyl, aryl

70 - 96 %

66 - 95 %
15 examples

Rapid oxidation of amino acids, coordinated to cobalt(III), to imines and amines by thionyl chloride in DMF

A. Hammershoi, R. M. Hartshorn, A. M. Sargeson, *J. Chem. Soc., Chem. Commun.*, 1988, (*18*), 1226-1227

62 %

87. Tin(II) chloride

CAS Registry Number	7772-99-8, 10025-69-1 (dihydrate)
CAS Name	Stannane, dichloro-
Molecular Formula	$SnCl_2$
Molecular Weight	189.60, 225.63 (dihydrate)
Boiling Point	652°C
Melting Point	246°C, 41-43°C (dihydrate)
Density	3.950 kg/m^3, 2.710 kg/m^3 (dihydrate)
Refractive Index	Not available.
Safety and Handling	Corrosive. Moisture sensitive.
Reactions	Allylation catalyst. Review: P. J. Smith, D. V. Sanghani, K. D. Bos, J. D. Donaldson, *Chem. Ind. (London)*, 1984, (5), 167-172.
Availability	Aldrich: 99.99+%, ££; anhydrous, 98%, p; dihydrate, 98%, ACS reagent, p; dihydrate, p.
	Lancaster Synthesis: dihydrate, p, bulk prices available.
	Johnson Matthey: dihydrate, crystalline, Specpure™, p.

Palladium-catalyzed carbonyl allylation by allylic alcohols with tin(II) chloride

Y. Masuyama*, J. P. Takahara, Y. Kurusa, *J. Am. Chem. Soc.*, 1988, **110**(*13*), 4473-4474

$$RCH=CHCH_2OH \ + \ RCHO \ \xrightarrow[\text{DMI, 25}^\circ]{\text{PdCl}_2\text{(PhCN)}_2 \, , \, \text{SnCl}_2} \ RCH=CHCH_2CH(OH)R$$

24 - 85 %
20 examples

Convenient synthesis of dimethyl selenide

M. G. Voronkov*, V. K. Stankevich, P. A. Podkuiko, N. A. Korchevin, E. N. Deryagina, B. A. Trofimov, *J. Gen. Chem. USSR*, 1987, **57**(*10,2*), 2144-2145

$$Se \ \xrightarrow[\text{ii) MeI, r.t.}]{\text{i) KOH, DMSO, SnCl}_2 , \, 90 - 95^\circ} \ SeMe_2$$

93 %
2 examples

Addition of acetals to activated olefins under extremely mild conditions

T. Mukaiyama*, K. Wariishi, Y. Saito, M. Hayashi, S. Kobayashi, *Chem. Lett.*, 1988, (*7*), 1101-1104

55 - 84 %
8 examples

88. Tin(IV) chloride
(Stannic chloride, tin tetrachloride)

CAS Registry Number 7646-78-8, 10026-06-9 (pentahydrate)

CAS Name Stannane, tetrachloro-

Molecular Formula $SnCl_4$

Molecular Weight 260.50, 350.58 (pentahydrate)

Boiling Point 114.1°C

Melting Point -33°C

Density 2.226 kg/m^3

Refractive Index 1.512

Safety and Handling Corrosive. Irritating to respiratory system, eyes and skin. Moisture sensitive.

Reactions Lewis acid catalyst.

Availability Aldrich: 99.999%, £; anhydrous, 99%, under N_2 in Sure/Seal™ bottles, p; 1M in CH_2Cl_2, under N_2 in Sure/Seal™ bottles, p; pentahydrate, 98+%, p.

Johnson Matthey: anhydrous, liquid (ampoule), Specpure®, p.

Aldol reaction with methylphenyl(oxoalkyl)benzothiazoline as an enolate transferring reagent

H. Chikashita*, S. Tame, K. Itoh, *Heterocycles*, 1988, **27**(*1*), 67-70

R^1= Ph, Pr
R^2= Ph, alkyl

+ R^2CHO $\xrightarrow[-78°, 6-12h]{SnCl_4 , CH_2Cl_2}$

15 - 67 %
5 examples

The asymmetric reaction of a chiral allylsilane with aldehydes

G.-L. Yi, D. Wang*, T. H. Chen, *Youji Huaxue*, 1988, **8**(2), 115-120

R = alkyl, Ph

+ RCHO $\xrightarrow[\text{ii) } H_2O \text{ or } H_2O/NaHCO_3]{\substack{\text{i) } SnCl_4 \text{ or } TiCl_4 , CH_2Cl_2 \\ -50 \text{ to } -30°, 20 - 40 \text{ min}}}$

56 - 90 %
1.2 - 14.7 % e.e.

Preparation of acyl isocyanates by zinc or tin catalyzed condensation of acyl chlorides with sodium cyanate

M.-Z. Deng, P. Caubere*, J. P. Senet, S. Lecolier, Tetrahedron, *1988*, **44**(*19*), 6079-6086

ArCOCl + NaOCN $\xrightarrow[180°, 2h]{o-Cl_2C_6H_4 , SnCl_4}$ ArCONCO

70 - 87 %
16 examples

89. Titanium(0)

CAS Registry Number	Not available.
CAS Name	Not available.
Molecular Formula	Not available.
Molecular Weight	Not available.
Boiling Point	Not available.
Melting Point	Not available.
Density	Not available.
Refractive Index	Not available.
Safety and Handling	Not available.
Reactions	Reduction. Reductive coupling of carbonyl compounds ("McMurry reaction"). Reviews: J. E. McMurry, *Acc. Chem. Res.*, 1983, **16**, 405-411; R. Dams, M. Malinowski, I. Westdorp, H. Y. Geise*, *J. Org. Chem.*, 1982, **47**, 248-259.
Availability	Not commercially available.
Preparation	Titanium(0) is generated *in situ* from reaction of $TiCl_4$ or $TiCl_3$ with strong reducing agents in THF.
Other Preparations	J. E. McMurry, *Acc. Chem. Res.*, 1974, **7**, 281; T. Mukaiyama, T. Sato, J. Hanna, *Chem. Lett.*, 1973, 1041; S. Tyrlik, I. Wolochowicz, *Bull. Soc. Chim. Fr.*, 1973, 2147.

Deoxygenation of halogen-containing heteroaromatic N-oxides

M. Malinowski*, L. Kaczmarek, *Synthesis*, 1987, (*11*), 1013-1015

$$R^1, R^2 = H, \text{aryl, benzo}$$

Ti(0) [TiCl$_4$/LiAlH$_4$], THF

0°, r.t., 15 min

84 - 98 %
11 examples

Preparation of diarylethylenediamines by aminative reductive coupling of benzaldehydes with low valency titanium reagents

C. Betschart, D. Seebach*, *Helv. Chim. Acta*, 1987, **70**(*8*), 2215-2231

HNR$_2$
R = alkyl

i) BuLi, hexane
ii) ArCHO
iii) TiCl$_4$, -70 to -10°, r.t.

iv) Ti(0) [TiCl$_4$/K or TiCl$_4$/Mg]
   -70 to r.t., 2 - 15 h
v) hydrolysis

23 - 81 %
22 examples

A convenient reduction of nitropyridine N-oxides to pyridinamines with titanium(0) reagent

M. Malinowski*, L. Kaczmarek, *J. Prakt. Chem.*, 1988, **330**(*1*), 154-158

Ti(0), [TiCl$_4$/Mg], THF
0°, r.t., 15 min

R = H, halo, alkyl, benzo

76 - 98 %
13 examples

90. Titanium(IV) chloride
(Titanic chloride, titanium tetrachloride)

CAS Registry Number	7550-45-0
CAS Name	Titanium chloride ($TiCl_4$)
Molecular Formula	$TiCl_4$
Molecular Weight	189.71
Boiling Point	136.4°C
Melting Point	-24°C
Density	1.730 kg/m^3
Refractive Index	1.61
Safety and Handling	Highly toxic. Corrosive. Irritating to eyes and respiratory system. Reacts violently with water.
Reactions	Reduction. Condensation. Lewis acid catalyst. Review: T. Mukaiyama, M. Murakami, *Croat. Chem. Acta*, 1986, **59**(*1*), 221-235.
Availability	Aldrich: 99.995+%, ££; 99.9%, under N_2 in Sure/Seal™ bottles, p; 1M in CH_2Cl_2, under N_2 in Sure/Seal™ bottles, p.

An efficient preparation of *N*-phosphinoyl and *N*-sulphonyl imines directly from aromatic aldehydes

W. B. Jennings*, C. J. Lovely, *Tetrahedron Lett.*, 1988, **29**(30), 3725-3728

$$PhCHO + Ph_2P(O)NH_2 \xrightarrow[\text{CH}_2\text{Cl}_2, 0°]{\text{TiCl}_4, \text{Et}_3\text{N}} \begin{array}{c} Ph \\ \diagup \\ H \end{array}{=}N{\diagdown}_{PPh_2} \quad 64\%$$

$$PhCHO + p\text{-TolSO}_2\text{NH}_2 \longrightarrow \begin{array}{c} Ph \\ \diagup \\ H \end{array}{=}N{\diagdown}_{SO_2pTol} \quad 81\%$$

12 examples

One-pot synthesis of allyl and alkyltrimethylsilanes

J. Pornet, A. Rayadh, L. Miginiac*, *Tetrahedron Lett.*, 1988, **29**(37), 4717-4718

$$\begin{array}{c} Me_3SiCH_2 \\ \diagup \\ H \end{array} \begin{array}{c} OEt \\ \diagdown \\ H \end{array} + MeCHO \xrightarrow[-70 - 0°]{\text{TiCl}_4, \text{CH}_2\text{Cl}_2} \begin{array}{c} Me_3SiCH_2 \\ \diagup \\ OHC \end{array} \begin{array}{c} Me \\ \diagdown \\ H \end{array}$$

50 %
6 examples

New synthesis of α-nitroso esters and α-keto ester oximes

S. M. Ali, Y. Matsuda, S. Tanimoto*, *Synthesis*, 1988, (*10*), 805-806

$$\begin{array}{c} Me \\ \diagdown \\ Me \diagup \end{array} \begin{array}{c} OEt \\ \diagdown \\ OSiMe_3 \end{array} \xrightarrow{\quad\quad} \begin{array}{c} Me \\ \diagdown \\ Me \diagup \end{array} \begin{array}{c} O \\ \| \\ C{-}OEt \\ NO \end{array} \quad \begin{array}{c} 68 - 75\% \\ 7 \text{ examples} \end{array}$$

i) TiCl$_4$, CH$_2$Cl$_2$
ii) NO or isoamyl nitrite

$$\begin{array}{c} n\text{-Bu} \\ \diagdown \\ \end{array} \begin{array}{c} OEt \\ \diagdown \\ OSiMe_3 \end{array} \xrightarrow{\quad\quad} \begin{array}{c} n\text{-Bu} \\ \diagdown \\ NOH \end{array} \begin{array}{c} O \\ \| \\ C{-}OEt \\ \end{array} \quad 65 - 70\%$$

91. Titanium(IV) isopropoxide (Tetraisopropyl orthotitanate)

CAS Registry Number 546-68-9

CAS Name 2-Propanol, titanium(4+) salt

Molecular Formula $(Me_2CHO)_4Ti$

Molecular Weight 284.26

Boiling Point $218°C/10$ mmHg

Melting Point $18-20°C$

Density 0.955 kg/m^3

Refractive Index 1.4654

Safety and Handling Flammable liquid. Moisture sensitive.

Reactions Catalyst for ring opening of epoxy alcohols and acids. Reviews: *Chem. Rev.*, 1961, **61**, 1; *Top. Cur. Chem.*, 1982, **106**, 3.

Availability Aldrich: p.

Lancaster Synthesis: 95%, p, bulk prices available.

Synthesis of α-substituted phenethylamines via titanium amide complexes

H. Takahashi*, T. Tsubuki, K. Higashiyama, *Synthesis*, 1988, (3), 238-240

$$\text{LiNR}^1_2 \xrightarrow[\substack{\text{ii) R}^2\text{CHO, 20}^\circ\text{, 3 - 4 h} \\ \text{iii) PhCH}_2\text{MgCl, Et}_2\text{O, r.t., 1 h} \\ \text{iv) H}_2\text{O}}]{\text{i) Ti(OPr-i)}_4\text{, Et}_2\text{O, -20}^\circ\text{, 20 min}}$$

R^1= alkyl
R^2= alkyl, aryl

Ph $\diagup$ R^2
NR$^1_2$

7 examples
73 - 93 %

Preparation of lactams via titanium(IV) isopropoxide mediated cyclization

M. Mader, P. Helquist, *Tetrahedron Lett.*, 1988, 29(25), 3049-3052

$$\text{RNH(CH}_2)_n\text{CHR'CO}_2\text{H} \xrightarrow[\text{under N}_2\text{, reflux, 3 - 26 h}]{\text{Ti(OPr-i)}_4\text{, Cl(CH}_2)_2\text{Cl}}$$

R, R' = H, Me
n = 1 - 3

O$\diagdown$ R'
RN — (CH$_2$)$_n$

35 - 93 %
6 examples

Mild and selective synthesis of homochiral *trans*-2,3-epithioalcohols from chiral *trans*-2,3-epoxyalcohols in the presence of titanium(IV) isopropoxide

Y. Gao, B. Sharpless*, *J. Org. Chem.*, 1988, 53(17), 4114-4116

n-C$_8$H$_{17}$ ... O ... OH $\xrightarrow[\substack{\text{ii) Ti(OPr-i)}_4\text{, THF, 25}^\circ\text{, 2 - 3 h} \\ \text{iii) NaHCO}_3}]{\text{i) NH}_2\text{CSNH}_2}$ n-C$_8$H$_{17}$ ... S ... OH

69 - 90 %
9 examples

92. Tributylstannyllithium

CAS Registry Number	17946-71-3
CAS Name	Lithium, (tributylstannyl)-
Molecular Formula	[Me(CH$_2$)$_3$]$_3$SnLi
Molecular Weight	Not available.
Boiling Point	Not available.
Melting Point	Not available.
Density	Not available.
Refractive Index	Not available.
Safety and Handling	Not available.
Reactions	Stannylation: W. C. Still, *J. Am. Chem. Soc.*, 1977, 99(*14*), 4836-4838. Reaction with α,β-enones: *J. Org. Chem.*, 1988, **53**(9), 1894-1899.
Availability	Not commercially available.
Preparation	*J. Organomet. Chem.*, 1968, **11**(2), 271-280.

Preparation of γ-ethoxyallylstannanes and dienol ethers

T. Takeda*, K. Ando, H. Ohshima, M. Inoue, T. Fujiwara, *Chem. Lett.*, 1986, (*3*), 345-348

54 - 86 %

84 - 97 %
5 examples

67 - 86%

60 - 86 %
6 examples

R¹, R² = alkyl

Desulphurizative stannylation of allyl sulphides using tributylstannyllithium

T. Takeda*, S. Ogawa, N. Ohta, T. Fujiwara, *Chem. Lett.*, 1987, (*10*), 1967-1970

R = alkyl, hetrocycle, Me₂NCS-
R¹- R³= H, alkyl

<88 %
100:0 - 0:100 mixtures
22 examples

One-pot synthesis of unsaturated macrolides, and substituted aromatics and heteroaromatics, via successive Michael reactions followed by ring closure annulation

G. H. Posner*, K. S. Webb, E. Asirvatham, S. Jew, A. Degl'Innocenti, *J. Am. Chem. Soc.*, 1988, **110**(*14*), 4754-4762

i) Bu₃SnLi
ii) CH₂=CHCOEt
iii) CH₂=CHCO₂Me
iv) Pb(OAc)₄

77 %
16 examples

93. Triethylborane

CAS Registry Number 97-94-9

CAS Name Borane, triethyl-

Molecular Formula Et_3B

Molecular Weight 98.00

Boiling Point $95^{o}C$

Melting Point $-93^{o}C$

Density 0.677 kg/m^3

Refractive Index 1.3971

Safety and Handling Toxic. Pyrophoric.

Reactions Catalyst in free radical reactions for addition to multiple bonds.

Availability Aldrich: under N_2 in Sure/Pac™ cylinders, p; 1M in hexanes, p; 1M in THF, p, both under N_2 in Sure/Seal™ bottles.

Synthesis of α-methylene-γ-butyrolactones

K. Nozaki, K. Oshima*, K. Utimoto, *Bull. Chem. Soc. Jpn.*, 1987, **60**(*9*), 3465-3467

R^1 = alkyl, aryl
R^2, R^3 = H, alkyl

70 - 84 %

31 - 59 %
5 examples

Free-radical substitution reactions for interconversion of alkenyl sulphides, -germanes, and -stannanes

Y. Ichinose, K. Oshima*, K. Utimoto, *Chem. Lett.*, 1988, (*4*), 669-672

R^1, R^2 = H, alkyl, aryl
X = SPh, SnPh$_3$

65 - 94 %

A convenient route to α-alkoxy esters from olefins through organoborane-catalyzed hydroalumination

K. Maruoka, K. Shinoda, H. Yamamoto*, *Synth. Commun.*, 1988, **18**(*10*), 1029-1033

i) Et$_3$B , AlCl$_2$H, hexane/CH$_2$Cl$_2$, 25°
ii) MeOClCHCO$_2$Me, reflux, 5 h
iii) dil. aq. HCl

$RR^1CHCH_2CH(OMe)CO_2Me$

43 - 70 %
4 examples

R, R^1 = H, alkyl

94. Trifluoroacetic anhydride

CAS Registry Number	407-25-0
CAS Name	Acetic acid, trifluoro-, anhydride
Molecular Formula	$(CF_3CO)_2O$
Molecular Weight	210.03
Boiling Point	39.5-40°C
Melting Point	-65°C
Density	1.487 kg/m^3
Refractive Index	not available
Safety and Handling	Corrosive. Harmful by inhalation. Moisture sensitive.
Reactions	Preparation of N- and O-trifluoroacetyl derivatives for GC analysis. Catalyst for esterification. Review: *Chem. Rev.*, 1955, **55**, 787.
Availability	Aldrich: 99+%, p.
	Lancaster Synthesis: 99+%, p, bulk prices available.
	Sigma: approx. 99%, p.

Acylation of aliphatic aldehyde hydrazones

Y. Kamitori, M. Hojo*, R. Msuda, T. Yoshida, S. Ohara, K. Yamada, N. Yoshikawa, *J. Org. Chem.*, 1988, 53(3), 519-526

$$\text{i-Pr}_2\text{N}-\text{N}=\text{CH}_2 \xrightarrow[\text{CHCl}_3\,,\,0°,\,1\,\text{min}]{(\text{F}_3\text{CCO})_2\text{O},\,2,6\text{-lutidine}} \text{i-Pr}_2\text{N}-\text{N}=\text{CHCOCF}_3$$

95 %
8 examples

86 %
5 examples

A one-pot synthesis of nitriles from alcohols

F. Camps*, V. Gasol, A. Guerrero, *Synth. Commun.*, 1988, 18(4), 445-452

$$\text{ROH} \xrightarrow[\text{ii) NaCN, HMPT, THF, reflux, 1 - 6.5 h}]{\text{i) (F}_3\text{CCO})_2\text{O, CH}_2\text{Cl}_2\,,\,\text{r.t., 15 min}} \text{RCN}$$

R = alkyl

85 - 99 %
9 examples

Ammonium nitrate/trifluoroacetic anhydride as a convenient reagent for N-nitration

C. Suri*, R. D. Chapman, *Synthesis*, 1988, (9), 743-745

78 %
4 examples

41 %
3 examples

95. Trifluoromethanesulphonic anhydride (Triflic anhydride)

CAS Registry Number   358-23-6

CAS Name   Methanesulfonic acid, trifluoro-, anhydride

Molecular Formula $(CF_3SO_2)_2O$

Molecular Weight   282.13

Boiling Point   81-83°C/745 mmHg

Melting Point   Not available.

Density   1.677 kg/m^3

Refractive Index   1.3212

Safety and Handling   Corrosive. Moisture sensitive.

Reactions   Catalyst for oxidation of alcohols by DMSO. Synthesis of alkyl and aryl triflates. Reviews: *Aldrichim. Acta*, 1983, **16**(*1*), 15; *Acc. Chem. Res.*, 1977, **10** 306.

Availability   Aldrich: ££.

Lancaster Synthesis: 98+%, ££, bulk prices available.

Sigma: ££.

Dehydration reactions with 'phosphonium anhydride' reagents

J. B. Hendrickson*, M. S. Hussoin, *J. Org. Chem.*, 1987, **52**(*18*), 4137-4139

Reaction 1: o-phenylenediamine (NH$_2$, NH$_2$) + PhCO$_2$H, Tf$_2$O/Ph$_3$PO, CH$_2$Cl$_2$, C$_2$H$_4$Cl$_2$, 0°; r.t., 30 min → 2-phenylbenzimidazole, 85 %, 7 examples

Reaction 2: Me—C$_6$H$_4$—CH=NOH, Tf$_2$O/Ph$_3$PO, CH$_2$Cl$_2$, C$_2$H$_4$Cl$_2$, r.t., 5 min → Me—C$_6$H$_4$—CN, 94 %, 4 examples

---

Synthesis of vinyl triflates using polymer-bound di-*tert*-butylpyridine

M. E. Wright*, S. R. Pulley, *J. Org. Chem.*, 1987, **52**(22), 5036-5037

cyclohexanone + polymer-bound di-tert-butylpyridine, Tf$_2$O, CCl$_4$, 25°, 24 h → cyclohexenyl OTf, 86 %, 3 examples

---

Synthesis of optically active *tert*-butylphenylphosphine sulphide, a source of new optically active organophosphorus compounds

Z. Skzypczynski, J. Michalski*, *J. Org. Chem.*, 1988, **53**(*19*), 4549-4551

t-Bu, Ph, S, P=S, OH, R-(+) → Tf$_2$O, CH$_2$Cl$_2$, -50° → t-Bu, Ph, P=S, OTf, R-(-) → H$_2$O, dioxane, 2 h, r.t. → t-Bu, Ph, P=S, OH, S-(-); → NaBH$_4$, EtOH → t-Bu, Ph, P=S, H, S-(-), 77 %

96. Trimethylsilyl trifluoromethanesulphonate (Trimethylsilyl triflate)

CAS Registry Number	27607-77-8
CAS Name	Methanesulfonic acid, trifluoro-, trimethylsilyl ester
Molecular Formula	$CF_3SO_3SiMe_3$
Molecular Weight	222.26
Boiling Point	$77°C/80$ mmHg, $39-40°/12$ mmHg
Melting Point	Not available.
Density	1.150 kg/m^3
Refractive Index	Not available.
Safety and Handling	Corrosive. Flammable. Very hygroscopic.
Reactions	Silylation. Review: *Aldrichim. Acta*, 1983, **16**(*1*), 15; 1984, **17**(*3*), 72; *Synthesis*, 1982, 1.
Availability	Aldrich: 99%, £. Lancaster Synthesis: 99%, £, bulk prices available. Sigma: Enquire for details.

Trimethylsilyl trifluoromethanesulphonate as activating agent for nucleophilic reactions between imines and Grignard reagents

M. A. Brook*, Jahangir, *Synth. Commun.*, 1988, **18**(*9*), 893-898

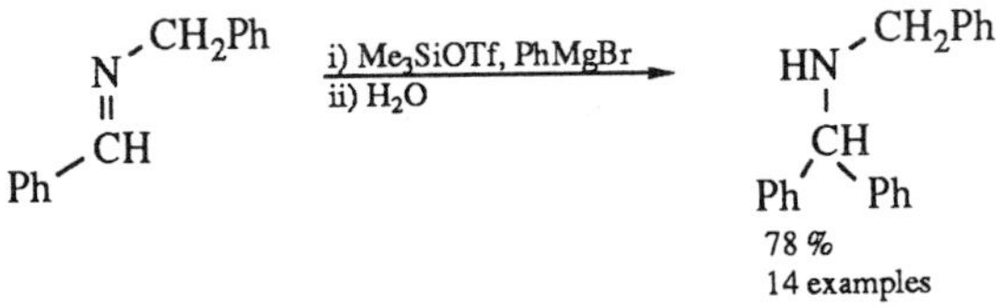

78 %
14 examples

Trimethylsilyl triflate-catalyzed aldol-type reactions of enol silyl ethers and acetals or related compounds

S. Murata, M. Suzuki, R. Noyori*, *Tetrahedron*, 1988, **44**(*13*), 4259-4275

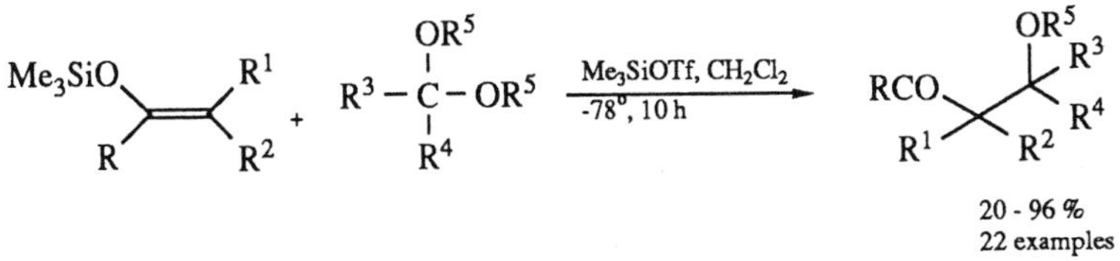

20 - 96 %
22 examples

New methods of β-conjugate addition and β-hydroxyalkylation of enones

S. Kim*, P. H. Lee, *Tetrahedron Lett.*, 1988, **29**(*42*), 5413-5416

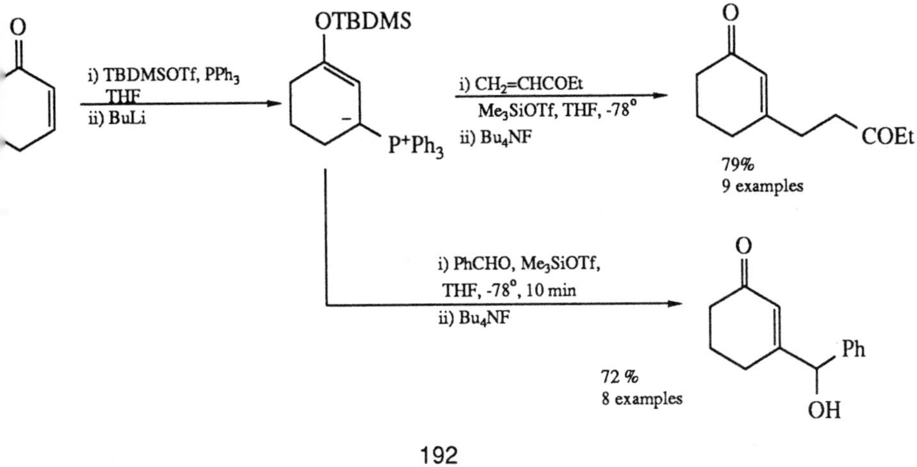

79%
9 examples

72 %
8 examples

97. Triphenylmethyl perchlorate (Trityl perchlorate)

CAS Registry Number	3058-33-1
CAS Name	Methylium, triphenyl-, perchlorate
Molecular Formula	Ph_3CClO_4
Molecular Weight	342.78
Boiling Point	Not available.
Melting Point	$143°C$
Density	Not available.
Refractive Index	Not available.
Safety and Handling	Potentially explosive.
Reactions	Addition catalyst. Cyclization catalyst. Review: T. Mukaiyama, M. Murakami, *Croat. Chem. Acta*, 1986, **59**(*1*), 221-235.
Availability	Not commercially available.
Preparation	Prepared by reaction of trityl chloride and silver perchlorate in nitrobenzene with precipitation on addition of benzene, or from trityl chloride or triphenylcarbinol in nitrobenzene or ether and perchloric acid followed by removal of all water.
Other Preparations	Improved preparation by Dauben *et al.*, *J. Org. Chem.*, 1960, **25**, 1442.

Facile synthesis of 1,5-dienes and β,γ-unsaturated nitriles via trityl perchlorate catalyzed addition

M. Murakami, T. Kato, T. Mukaiyama, *Chem. Lett.*, 1987, (6), 1167-1170

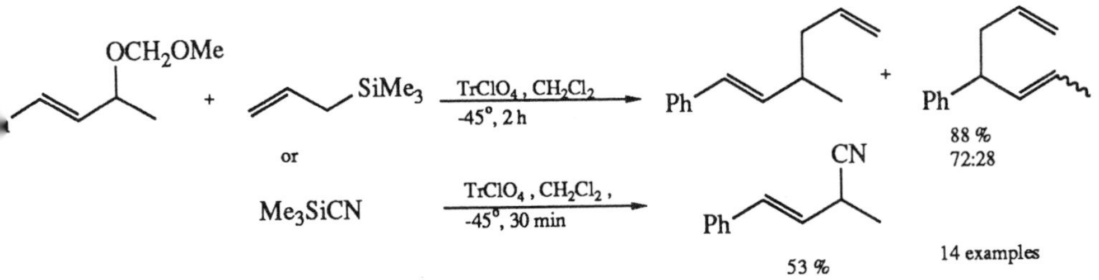

88 %
72:28

53 %

14 examples

Trityl cation catalyzed cyclization of alkoxyalkyl- and alkoxyalkenylsilanes

Y.-L. Chen, T. J. Barton*, *Organometallics*, 1987, 6(12), 2590-2592

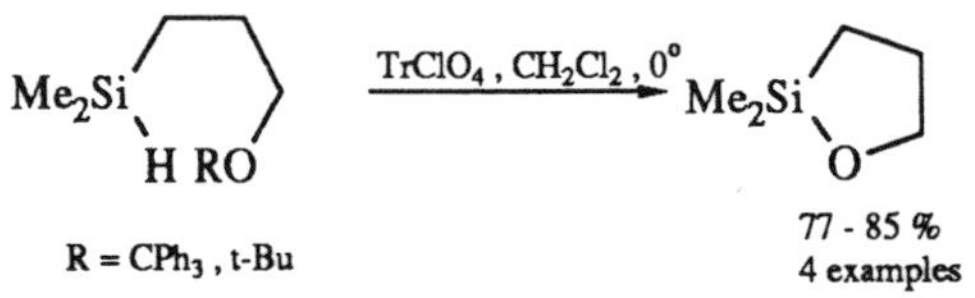

R = CPh₃ , t-Bu

77 - 85 %
4 examples

Regioselective addition of 1,2-benzoquinones to silyl enol ethers catalyzed by trityl perchlorate

Y. Sagawa*, S. Kobayashi, T. Mukaiyama, *Chem. Lett.*, 1988, (7), 1105-1108

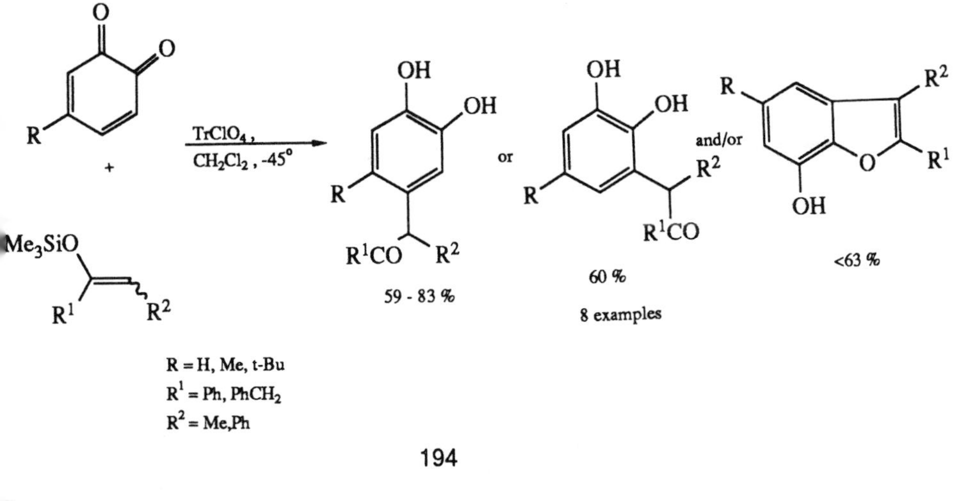

59 - 83 %

60 %

8 examples

<63 %

R = H, Me, t-Bu
R¹ = Ph, PhCH₂
R² = Me, Ph

98. Triphenylphosphine

CAS Registry Number	603-35-0
CAS Name	Phosphine, triphenyl-
Molecular Formula	Ph$_3$P
Molecular Weight	262.29
Boiling Point	377°C
Melting Point	79-81°C
Density	1.0749 kg/m^3
Refractive Index	1.6358
Safety and Handling	Irritant.
Reactions	Versatile reducing agent. Reagent for preparation of triphenylphosphonium salts (Wittig). Review: *Organophosphorus Reagents in Organic Synthesis*, ed. J. I. G. Cadogan, Academic Press, 1979.
Availability	Aldrich: 99%, p; polymer-supported (polystyrene + DVB), £££. Also available as borane complex.
	Lancaster Synthesis: 99%, p, bulk prices available.
	Sigma: crystalline, p.

Synthesis of isoquinolines by condensation of iminophosphoranes with o-phthalaldehyde

T. Aubert, M. Farnier, B. Hanquet, R. Guilard*, *Synth. Commun.*, 1987, **17**(*15*), 1831-1837

$$MeCOCH_2N_3 \xrightarrow[\substack{\text{i) PPh}_3, CH_2Cl_2, 0° \\ \text{ii) o-C}_6H_4(CHO)_2, CH_2Cl_2 \\ 0°, 0.5 h; r.t. overnight}]{}$$

COMe

30 %
6 examples

Regiospecific formation of dienes by the palladium-catalyzed inter- and intramolecular coupling of vinyl halides

R. Grigg*, P. Stevenson, T. Worakun, *Tetrahedron*, 1988, **44**(*7*), 2049-2054

Pd(OAc)$_2$, PPh$_3$, K$_2$CO$_3$, MeCN
reflux, 4.5 h

90 %
6 examples

Novel phosphorane and phosphonate synthons for vinyl glycines

A. J. Bicknell, G. Burton, J. S. Elder*, *Tetrahedron Lett.*, 1988, **29**(*27*), 3361-3364

$$Br \xrightarrow{CO_2Et} \xrightarrow[\substack{\text{MeONH}_2.HCl \\ EtOH, PPh_3, THF \\ reflux}]{} Ph_3P^+ \xrightarrow{CO_2Et} \xrightarrow[\substack{\text{i) K}_2CO_3, DMF \\ \text{ii) RCOR'} \\ \text{iii) Zn, HCO}_2H}]{} \xrightarrow[R']{R} \xrightarrow[NH_2]{CO_2Et}$$

R = alkyl, aryl
R'= H, alkyl
RR'= cycloalkyl

37 - 99 %
6 examples

99. Xenon difluoride

CAS Registry Number	13709-36-9
CAS Name	Xenon fluoride (XeF_2)
Molecular Formula	XeF_2
Molecular Weight	169.29
Boiling Point	Not available.
Melting Point	~140°C
Density	Not available.
Refractive Index	Not available.
Safety and Handling	Not available.
Reactions	Fluorination.
Availability	Not commercially available.
Preparation	J. L. Weeks, M. S. Matheson, *Inorg. Synth.*, 1966, **8**, 260-264.

Fluorination of 1,3-dienes with xenon difluoride and (difluoroiodo)benzene

D. F. Shellhamer*, R. J. Conner, R. E. Richardson, V. L. Heasley, G. E. Heasley, *J. Org. Chem.*, 1984, **49**(*25*), 5015-5018

$$\underset{H_2C}{\overset{Me}{\diagdown}}\overset{Me}{\underset{CH_2}{\diagup}} \xrightarrow{\text{XeF}_2 , \text{BF}_3.\text{Et}_2\text{O}, 25°, 20\text{ min}} FCH_2 - \underset{F}{\overset{Me}{\underset{|}{\overset{|}{C}}}} - \overset{Me}{\underset{|}{C}} = CH_2$$

80 %
4 examples

Replacement of the carboxylic acid function with fluorine

T. B. Patrick*, K. K. Johni, D. H. White, W. S. Bertrand, R. Mokhtar, M. R. Kilbourn, M. J. Welch, *Can. J. Chem.*, 1986, **64**(*1*), 138-141

$$RCO_2H \xrightarrow{\text{XeF}_2 , \text{CH}_2\text{Cl}_2, 22°, 10\text{ h}} RF$$

R = alkyl, aryl

3 - 95 %
21 examples

Direct perfluoroalkylation of aromatic compounds using perfluorocarboxylic acids and xenon difluoride

Y. Tanabe, N. Matsuo*, N. Ohus, *J. Org. Chem.*, 1988, **53**(*19*), 4582-4585

$$\xrightarrow[\text{CH}_2\text{Cl}_2 , \text{r.t.}]{\text{XeF}_2 , \text{CF}_3\text{CO}_2\text{H},}$$

72 %
16 examples

100. Zinc borohydride

CAS Registry Number   17611-70-0

CAS Name   Borate(1-), tetrahydro-, zinc (2:1)

Molecular Formula $Zn(BH_4)_2$

Molecular Weight   95.06

Boiling Point   Not available.

Melting Point   Not available.

Density   Not available.

Refractive Index   Not available.

Safety and Handling   Not available.

Reactions   Reduction: *Chem. Pharm. Bull.*, 1984, **32**(*4*), 1411-1415.

Availability   Not commercially available.

Preparation   Prepared from $ZnCl_2$ and $NaBH_4$ in anhydrous ether: *J. Am. Chem. Soc.*, 1960, **82**, 6074.

Other Preparations   S. Kedrova, V. N. Konoplev, N. N. Mal'tseva, L. N. Tolmacheva, N. S. Kurnakov, *Otkrytiya Izobret., Prom. Obraztsy, Tovarnye Znaki*, 1975, **52**(*48*), 64; V. I. Mikheeva, N. N. Mal'tseva, N. S. Kedrova, *Zh. Neorg. Khim.*, 1979, **24**(*2*), 408-413.

Mild reduction of indoles to indolines with zinc borohydride

H. Kotsuki*, U. Ushio, M. Ochi, *Heterocycles*, 1987, **26**(*7*), 1771-1774

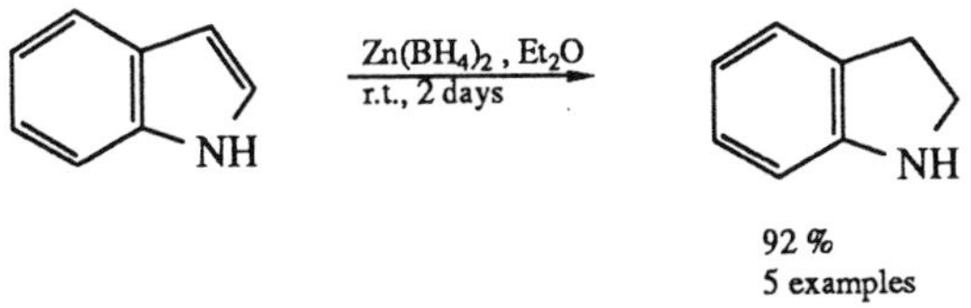

$Zn(BH_4)_2$, Et_2O
r.t., 2 days
NH
NH
92 %
5 examples

New and convenient synthesis of dihydrobenzisoquinolinones using sodium or zinc borohydride

R. Sato*, K. Oikawa, T. Goto, M. Saito, *Bull. Chem. Soc. Jpn.*, 1988, **61**(*6*), 2238-2240

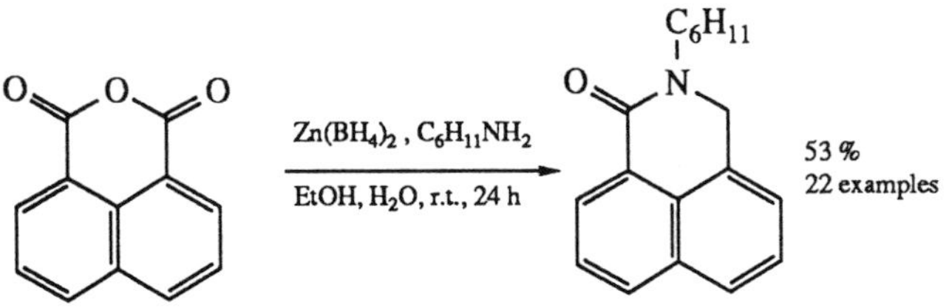

$Zn(BH_4)_2$, $C_6H_{11}NH_2$
EtOH, H_2O, r.t., 24 h
C_6H_{11}
53 %
22 examples

Efficient reduction of acyl chlorides with zinc borohydride and tetramethylethylenediamine

H. Kotsuki*, Y. Ushio, N. Yoshimura, M. Ochi, *Bull. Chem. Soc. Jpn.*, 1988, **61**(*7*), 2684-2686

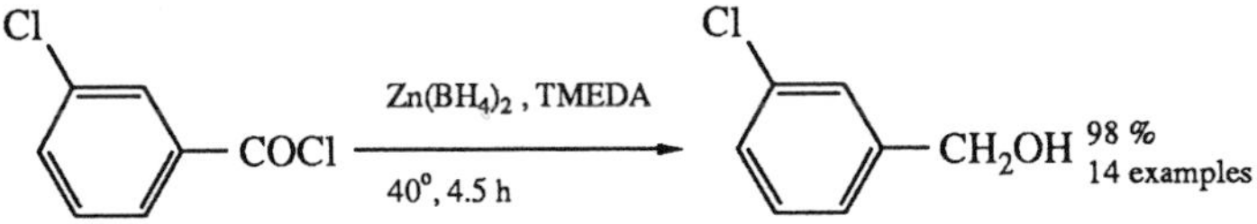

Cl
COCl
$Zn(BH_4)_2$, TMEDA
40°, 4.5 h
Cl
CH_2OH
98 %
14 examples

Reaction Index

Reaction Index

**The numbers in the index refer to the reagent numbers
and *not* to page numbers**

Looking for novel or unusual reactions and synthetic methods relevant to your research interests?

Don't let your time slip through the hourglass of manual literature searching! Instead, read:

METHODS IN ORGANIC SYNTHESIS

METHODS IN ORGANIC SYNTHESIS (MOS) is a well established and respected current awareness publication for workers in the field of organic synthesis.
Each monthly issue contains over 200 items which include titles, bibliographic details and reaction schemes. For example:

10425   Synthesis of vinyl triflates using polymer-bound di-*tert*-butylpyridine
M. E. Wright*, S. R. Pulley   *J. Org. Chem.*, 1987, **52**, (22), 5036-5037

MOS also has five indexes, making it quick and easy to scan.